MATHEMATICS RESEARCH DEVELOPMENTS

CHAOS AND COMPLEXITY IN THE ARTS AND ARCHITECTURE

RESEARCH IN PROGRESS

MATHEMATICS RESEARCH DEVELOPMENTS

Additional books in this series can be found on Nova's website under the Series tab.

Additional e-books in this series can be found on Nova's website under the eBooks tab.

CHAOS AND COMPLEXITY IN THE ARTS AND ARCHITECTURE

RESEARCH IN PROGRESS

NICOLETTA SALA
AND
GABRIELE CAPPELLATO
EDITORS

nova
science publishers
New York

Copyright © 2018 by Nova Science Publishers, Inc.

NOTICE TO THE READER

Library of Congress Cataloging-in-Publication Data

ISBN: 978-1-53612-995-3

Published by Nova Science Publishers, Inc. † New York

CONTENTS

PREFACE

This book is organized in seventeen chapters, and is divided in two logical parts. First part, Chapter 1 until Chapter 10, introduces new researches in the fields of the connections between fractals, arts and architecture. Second part, from Chapter 11 to Chapter 17, introduces new processes of generation fractal shapes, which can be used in the arts.

Chapter 1, Gabriele Cappellato and Nicoletta Sala introduce the fractality in the arts and architecture. The authors describe as fractal geometry, in particular its property of self-similarity and the processes of bifurcation are present in different epoch, and in different cultures around the world. The examples presented in the chapter, help to deduce which fractal geometry and the self-similarity can be thought as a code of the forms.

Chapter 2, Vincenzo Iorfida, Mauro Francaviglia, and Marcella Giulia Lorenzi, present as curves and varieties have acted as sources of inspiration for artistic themes in many of the "geometrical forms" of Modern Art, in 19th and 20th Centuries. Nowadays, their beautiful shapes can be easily constructed by computers. The authors shortly discuss their role, providing a few examples from Painting to Sculpture and Architecture.

Chapter 3, Vincenzo Iorfida, describes as the 20th century represents a moment of total change for Science, as Renaissance for art. Science and its new counterintuitive models are synchronized with a creative art form dedicated to spread the difference between actual reality and the detected one. Both realities propose again the ancient dilemma of the dichotomy "continuous - discrete". In the chapter, through quantum nanotechnologies evolution, the authors approaches an art world devoted to the entanglement phenomenon modifying the perceptual conception of space, and offering new horizons in the knowledge transfer in educational and divulgation fields.

Chapter 4, Arturo Buscarino, Luigi Fortuna, Mattia Frasca, Angelo Lamia, and Maria Gabriella investigate the role and the characteristics of fog in various paintings by famous artists of different art movements, in which the presence of fog significantly affect the visual experience. To do this a new nonlinear signal processing technique, able to remove fog from color pictures exploiting optical properties of fog effect is introduced and its implementation on a Cellular Nonlinear Network (CNN) is described. Based on the assumption that if fog is a real element of the natural scenario then the artist can catch its optical effect in the artwork, the proposed methodology is used to investigate whether the fog is a natural element or it has been artificially added by the artist to express its own feelings. A further analysis has been carried on to establish if the fog in some particular paintings may play the same role of noise in stochastic resonance allowing to enhance some features which cannot be distinguished in absence of fog.

Chapter 5, Nicoletta Sala describes how, during the centuries, the architecture has followed the Euclidean geometry and Euclidean shapes. Thus, the buildings had Euclidean aspects, but some architectural styles, for example the Baroque and the Hindu architecture, are informed by Nature, and much of Nature is manifestly fractal and complex. The author describes where the fractality appears in architecture and in urban organization, opening new opportunities in the virtual architecture and in the hyperarchitecture, too.

Chapter 6, Nicoletta Sala analyses the connection between complexity and architecture. She starts observing that architecture finds inspiration observing the nature, and nature is fractal and complex. Modern architects study the complexity and the fractal geometry to create a new kind of buildings or to understand the problems connected to the networks' organizations and to the urban growths. The aim is to present an approach that studies the complexity applied in architecture.

Chapter 7, Mamta Rani, Sanjaya Tripathi, and Arun Prakash Agarwal show that germs of fractals exist in old Indian literature, e.g., fractal architecture in Indian temples and fractal weapons. The purpose of this paper is to collect a few examples from old Indian history and present their fractal aspects.

Chapter 8, Nicoletta Sala starts from the point of view that fractal geometry is an excellent tool for modelling natural shapes (e.g., textures, ferns, trees, flowers, seashells, rivers, mountains), and its important applications appear in computer science. In particular, fractal geometry permits to reproduce, in computer graphics and in the virtual reality, the complex and irregular forms present in nature using simple iterative or recursive instructions. The author presents some recent applications for generating virtual landscapes, territories and complex shapes in virtual worlds and in Internet based virtual worlds, for example Second Life.

Chapter 9, Nicoletta Sala presents some examples of industrial design objects analysed using the complexity and the fractal geometry. Complex and fractal components appeared in the industrial design after the development of materials, for example the introduction of the float glass, and the manufacturing techniques, for example the work with the laser. This exploration and development of materials, manufacturing techniques and design are often indistinguishable from one another.

Chapter 10, Nicoletta Sala describes how the concept of time could be applied in architecture. Few architects have considered the role, which the time has on their buildings. For example, the time and the light could break the symmetry of an architectonic design, creating sensations of caducity, or new ties, which the architects have not thought. The time and the atmospheric agents corrode and damage the buildings modifying many architectures, and the nature has worked on them, creating new hybrid forms. The chapter introduces some examples of contemporary projects, where the architects introduce the dynamic in the buildings' shapes.

Chapter 11, Robert A.M. Gregson, introduces the Synchronization of Fractals in Logarithmic Spirals. Synchronization is not treated here is a fundamental necessary property, but is transient and derivative on sequences that are generated by nonlinearity in time series which arise in neural brain processes, and in bottom-up top-down network dynamics. The author finds both symmetrical and spiral patterns on local regions of Julia sets, and discontinuous series in the dynamics of some region that are recordable in the neurophysiology of intermittent consciousness. Synchronization can also be called self-similarity, in induced noncommutative geometry.

Chapter 12, Mamta Rani, Riaz UlHaq, and Norrozila Sulaiman present new Koch curves generated by dividing the initiator into unequal parts. With the increase in size of the set of Koch curve, we felt that there is a need of classification of Koch family. The classification is based on their method of generation.

Chapter 13, Manish Kumar and Mamta Rani describes new Sierpinski Curve Julia Sets.

Chapter 14, Saurabh Goel and Mamta Rani presents the Hilbert curve. It is one of the space-filling curves generated by dividing the initiator into equal parts. In literature, many superior fractals have been created by dividing the initiator into unequal parts. In this paper, we have enriched the gallery of superior fractals by adding new Hilbert curves into it. Also, production rules to draw the new Hilbert curves have been developed. Further, it is interesting to see the new production rules for the conventional Hilbert curve also.

Chapter 15, Mamta Rani, Deepak K. Verma, and J. S. Sodhi, present new beautiful superior Julia sets have been generated for z^n+c, n ≥ 4 in superior orbit, and modeled in 3 dimensions. Interesting shapes which remind artistic works of computer art.

Chapter 16, Mamta Rani, and Bharti Singh, describe the Ginko Leaves. Ginko is a "living fossil", and has been declared as "tree of the millennium". The authors show that there are many ways to generate Gingko Gingko leaf.

Chapter 17, Mamta Rani, R. C. Dimri, and Darshana J. Prajapati, present the V-variable fractals and superfractals, which have been recently introduced by Barnsley, Hutchinson, and Stenflo to the fractal graphics world. Superior iterates have emerged as a new powerful tool in the study of discrete dynamics and fractal theory. The authors have developed techniques to generate Sierpinski Gasket and Sierpinski Carpet as 3-variable and as 4-variable fractals respectively using superior iterates for contractive operators.

INTRODUCTION

The Mother Art is Architecture. Without an architecture of our own we have
no soul of our own civilization.

Frank Lloyd Wright

Mathematics and geometry are basic elements for composing an architecture project. They are science of space of relations and proportions because they deal with the qualities and properties of shapes: in fact, applying two-dimensional graphic operations, geometry is capable of building three-dimensional and therefore spatial shapes.

Mathematics and geometry are the disciplinary tools for the architect in treating quantities for building volumes and structures in space.

Mathematics and geometry belong to the architecture of any time, they are the guiding and hidden order of any project. They help the perfection and proportion between the various parts of the architectural design, as the trace of the different forms in the evolution of the work to its realization.

Correct and rigorous architecture develops through a geometrical orientation principle by interacting simple solid figures with complex shapes, and design orders in apparent building freedom. It is within the architecture with the same intensity with which mathematics interacts with the natural proportional construction of nature. The relationship between geometry, mathematics and architecture is expressed within the design process and the consequences of these three disciplines in the execution of the project.

Architecture has the origin of the principles governing nature. For example, the concept of gravity, vertical lines such as the trees that simulate the pillars; the horizontal line indicating the protection and gives rise to coverage; the right angle that outlines a portion of territory or the curved line of the horizon that suggests the circle and strengthens the idea of a volume.

These are all elements that reconstruct the archetypal idea of geometry as a principle of architecture in nature, while the later evolution leads to the knowledge of more rigorous scholars and mathematicians such as Pythagoras, Talete, Euclid.

The history of architecture is full of examples where geometry is understood as the image of a natural landscape (in Figure 1 an example). For example, the peasant's house is built through elementary geometric shapes that contrast with nature; with the same force the volume of the castle or that of the more complex monastery is highlighted in the landscape with their shapes obtained by the composition of different geometric and articulated solids.

Even in urban nuclei, geometry is solved through simple and immediate architectural constructions. A view from above shows clearly in the evolution of the city the basic connective tissue consisting of full forms and empty geometries interacting proportionally with each other, resulting in the aggregation of any urban nucleus until reaching the contemporary city where the skyscraper becomes the symbol which emerges in the thick texture of the metropolitan building. In this case, geometry is expressed as a principle of iconic values, exalting the quality of the work and charging it with symbolic public values in which political, religious, cultural institutions come into play.

The use of geometry as a principle of form is also intended to express the contents of the architectural work and is particularly useful in the period of the Enlightenment (18th century), where the interest in elementary geometric form was a profoundly revolutionary attitude. We think to Étienne-Louis Boullée, Claude-Nicolas Ledoux, Jean-Jacques Léqueu, all architects that have shown a break in their designs over the traditional patterns of that time (an example of Boullée's work is in Figure 2). Spheres, cylinders, cones, pyramids, cubes, parallelepipeds, elemental and symbolic volumes are being experimented in the idea of a project as they are thought of as the principle of form; this type of geometry is also related to the spatial qualities that geometric solids themselves can express in these projects.

The symbolic qualities and the identity of a project can be derived from the choice of geometry of elementary figures as the absolute factor of the project system. Thus, geometry and mathematics are no longer a principle, but they become a control tool for the elaboration of any good architectural project where the relationship between composition and geometry needs to be balanced so that there is a correct correspondence between the spatial qualities of the environments and geometric characteristics of the figures that determine them. Only knowing the universe of mathematics and geometry it is possible to understand the proportional structure and internal properties of each figure, to both technically and from viewpoint of perception and meaning the whole work that will be built.

Figure 1. The origin of the Corinthian Order, found inspiration by the nature (By Unknown - Claude Perrault's French translation of Vitruvius (1684),Transferred from en.wikipedia to Commons by MARKELLOS using CommonsHelper., Public Domain, https://commons.wikimedia.org/w/index.php?curid=17805683).

Figure 2. Cénotaphe à Newton (1784) by Boullée (By Étienne-Louis Boullée - Own scan from: "Klassizismus und Romantik. 1750-1848", Hrsg. Rolf Toman, Verlag Ullmann und Könemann, Sonderausgabe, ISBN 978-3-8331-3555-2, Public domain, https://commons.wikimedia.org/w/index.php?curid=2139341, {{PD-1923}} – published anywhere before 1923 and public domain in the U.S.).

Contemporary architecture and art can find inspiration in the theory of chaos and complexity. For example, we think of a painting by Pollock that has, through his art, anticipated the theory of complexity, or some buildings, such as Santiago Calatrava'sGare de Oriente (1998, Lisbon, Portugal) where bold and sinuous forms, seem to come out of graphs of complex functions, in Figure 3.This is one of those we have called "architectures of the complexity".

Figure 3. Santiago Calatrava's Gare de Oriente (1998, Lisbon, Portugal) (By Martín Gómez Tagle - Lisbon, Portugal, CC BY-SA 3.0, https://commons.wikimedia.org/w/index.php?curid=13764903).

These unusual forms have been realized using new CAD / CAE systems, in which it is possible to analyze the structural elements in their interaction and complexity. In order to reach them, in the last century, it has been through different movements: from futurism, to neoplasticism, to organic architecture, finally to deconstructivism. Fractal geometry in this field plays a primary role in explaining and justifying the forms that vanguard architects often use.

Quoting Swedish architect Alvar Aalto:

> The ultimate goal of the architect...is to create a paradise. Every house, every product of architecture... should be a fruit of our endeavour to build an earthly paradise for people.

We think that chaos and complexity, with their hidden order, can help the evolution of the arts and architecture. In particular, they could help the architecture for realizing this paradise.

Nicoletta Sala
Institute for Scientific and Interdisciplinary Studies (ISSI),
Locarno, Switzerland
Email: nicoletta.sala@usi.ch.
and
Gabriele Cappellato
Academy of Architecture, Mendrisio, Switzerland
Email: gabriele.cappellato@usi.ch.

In: Chaos and Complexity in the Arts …   ISBN: 978-1-53612-995-3
Editors: N. Sala and G. Cappellato   © 2018 Nova Science Publishers, Inc.

Chapter 1

FRACTALITY IN THE ARTS
AND IN ARCHITECTURE

Gabriele Cappellato[1,] and Nicoletta Sala[2,3,4,†]*

[1]Accademia di Architettura di Mendrisio, University of Lugano,
Lugano, Switzerland
[2]Institute for Complexity Study, Rome, Italy
[3]CERFIM (Research Center for Mathematics and Physics),
Locarno, Switzerland
[4]ISSI (Institute for Scientific and Interdisciplinary Studies),
Locarno, Switzerland

ABSTRACT

The aim of this work is to describe as fractal geometry, its property of self-similarity, and its processes of bifurcation can be appear in the arts, and in architecture. These fractal features are common in different cultures and in different architectural styles.

Keywords: arts, architecture, bifurcations, fractal geometry, fractality, self-similarity

[*] E-mail: gabriele.cappellato@usi.ch.
[†] E-mail: nicolettasala@virgilio.it (Corresponding author).

1. INTRODUCTION

Fractal geometry is a modern mathematical theory that radically departs from traditional Euclidean geometry, which has been applied for more than two millennia to interpret the nature and its phenomena.

Fractal geometry describes objects that are scale symmetric, or self-similar. This means that when such objects are magnified, their parts are seen as an exact resemblance to the whole, the property continues with the parts of the parts and so on to infinity.

These shapes are called "fractals," and they must maintain a rough, jagged quality at every scale at which an object can be examined. The aim of this work is to present how fractal geometry, in particular the property of self-similarity and the process of bifurcation, can find application in the arts, and in architecture (as described in section 2, and in section 3 respectively).

1.1. Fractality in the Arts

Across the centuries, many artists have experienced the fascination of fractal geometry. In particular, the property of self-similarity is in fact common in different cultures.

In the Western culture, we can find self-similar forms in the "Cosmatesque" mosaic floor of *Santa Maria in Trastevere* (Rome), *San Lorenzo Fuori le Mura* (Rome), *Cathedral of Civita Castellana* (Italy). These mosaics are in Cosmatesque style, which is a definition used in the history of art and architecture, with regard to a type of ornamentation characteristic of the Roman mosaics of marble in the 12th and 13th century, but of Byzantine origin. It consisted of sophisticated and imaginative shapes. In the 19th century, this decoration was called "Cosmatesque." This term was coined, by Italian architect Camillo Boito (1836-1914), in his book *Architettura Cosmatesca (Cosmatesque Architecture)* (1860, p. 4). It derived by the Cosmati family, so called by a member, Cosimo or Cosma. The Cosmati family was a Roman family where the members were architects, sculptors and workers in decorative geometric mosaic.

Figure 1 shows three examples of Cosmatesque motifs. We can observe the presence of self-similar triangles and the Sierpinski triangles. In Figure 2 there is an example of Cosmatesque pavement. It is inside the church of *San Benedetto in Piscinula* (Rome, Italy). Other examples of Cosmatesque mosaics are in *Santa Maria in Cosmedin* (782-1123, Rome), in *Santa Maria Maggiore*

(1185-1210, Civita Castellana, Italy). Figure 3 shows a collection of Cosmatesque mosaics.

Another interesting example of Cosmatesque mosaic is in the *Cattedrale di Santa Maria SS Annunziata* (*Cathedral of Assumption of Mary*, 1072-1104) located in Anagni (Frosinone, Italy). Inside the cathedral, there is a floor, a remarkable mosaic floor realized in 1231. The floor is adorned with dozens of mosaics, each in the form of a Sierpinski fractal gasket, but it impressive the analogy with an Apollonian gasket. The Apollonian gasket corresponds to a limit set that is invariant under a Kleinian group. Kleinian group is a finitely generated discontinuous group of linear fractional transformations $z \rightarrow (az + b)/(cz+d)$ acting on a domain in the complex plane.

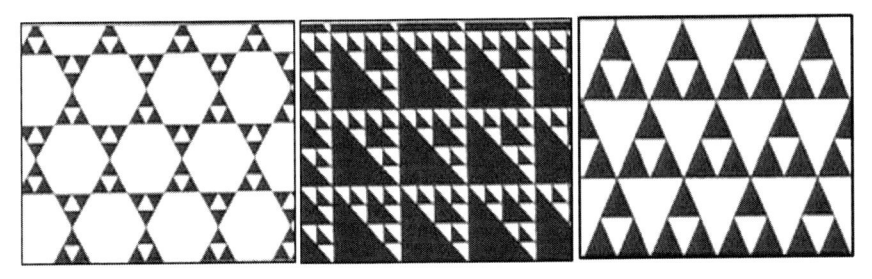

Figure 1. Cosmatesques motifs: three examples where the Sierpinski triangle is present.

Figure 2. Self-similar floor decorations: Cosmatesque pavements in the church of *San Benedetto in Piscinula* (Rome, Italy) (By Lalupa – Own work, CC BY-SA 3.0, https://commons.wikimedia.org/w/index.php?curid=2400683).

Figure 3. Cosmatesque mosaics (http://artesplorando.blogspot.it/2015/07/il-mosaico-2-da-bisanzo-lucio-fontana.html).

In African cultures, we can find fractal components. Ron Eglash, in his book *African Fractal Modern Computing and Indigenous Design* (1999), investigates fractals in African arts and architecture. He also analyzes the social and political implications on the existence of African fractal geometry. On the use of fractal shapes in an African artisan artwork, Eglash introduces the fractal esthetic as "the artisan's intuition or sense of beauty" (Eglash, 1999, p. 52). Three examples of African sculptures are in Figures 4a, 4b and 4c, where the presence of the self-similarity in the evident.

We can observe the same fractal property in multiple heads statues of Buddhist God (examples in Figures 4d and 4e).

In the Japanese arts we can find the presence of the self-similar spirals and the process of bifurcation in the Hokusai's woks. Hokusai's full name was Katsushika Hokusai (1760-1849), Japanese painter and wood engraver, born in Edo (now Tokyo). He is considered one of the outstanding figures of the Ukiyo-e, or "pictures of the floating world" (everyday life), school of printmaking. Hokusai entered in the studio of his countryman Katsukawa Shunsho in 1775 and there learned the new, popular technique of woodcut printmaking. Between 1796 and 1802 he produced a vast number of book illustrations and colour prints (perhaps as many as 30,000) that drew their inspiration from the traditions, legends, and lives of the Japanese people. Hokusai's most typical wood-block prints, silk-screens, and landscape paintings were done between 1830 and 1840. The curved lines characteristic of his style gradually developed into a series of spirals that imparted the utmost freedom and grace to his work, as in *Raiden, the Spirit of Thunder*. In *Amida falls* (1834-1835), shown in Figure 5a, Hokusai represents the falls as a sub-harmonic function illustrates in Figure 5b (Fivaz, 1988). In the Hokusai's *Kirifuri Waterfall at Mount Kurokami in Shimotsuke province* (1832), Nelson-Atkins Museum of Art, Kansas City (Missouri), shown in Figure 5c, there is a realistic depiction of Kirifuri waterfall, one of the three famous waterfalls of Nikko (Japan). Particularly impressive is the analogy of a fractal process of bifurcation (Sala, 2004).

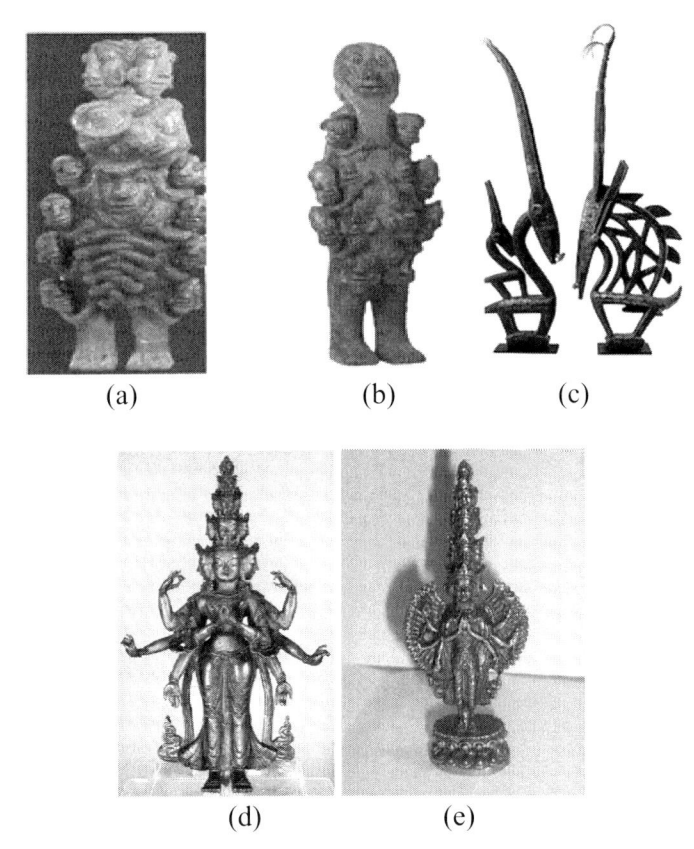

(a) (b) (c)

(d) (e)

Figure 4. (a,b,c) Self-similarity in African and (d, e) oriental art.

Figure 6 illustrates another work by Hokusai: *The Breaking Wave off Kanagawa*, also called *The Great Wave* (1831). Woodblock print from Hokusai's series *Thirty-six Views of Fuji*, which are the high point of Japanese prints. The original is at the Hakone Museum in Japan. In *The Great Wave*, there are three boats among the turbulent, broken waves. The boats mould into the shapes of the engulfing waves. Figure 6a shows this Hokusai's work, and in the Figure 6b there is real wave, with its complexity. Observing the Figure 6a, we can note the presence of some different self-similar spirals.

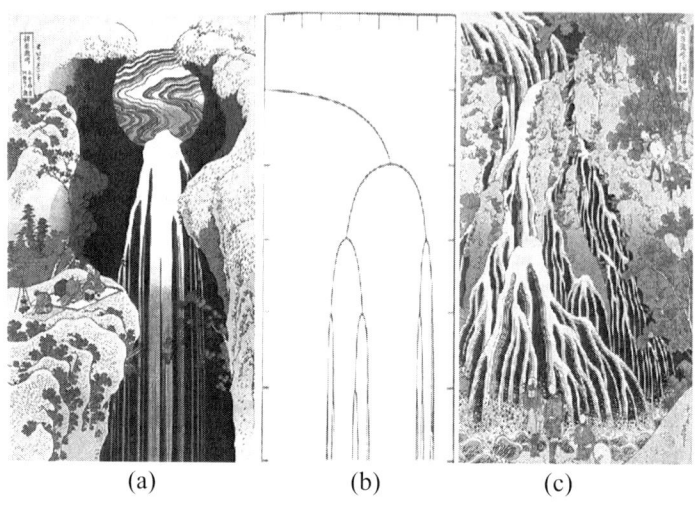

(a)	(b)	(c)

Figure 5. (a) Hokusai's *Amida Falls* (1834-1835) and (b) the sub-harmonic function (c) Hokusai's *Kirifuri Waterfall at Mount Kurokami in Shimotsuke province* (1832).

(a)	(b)

Figure 6. (a) *The Great Wave Off Kanagawa* (1831) by Hokusai, it has self-similar spirals. (b) A real wave and its dynamic and complex shape.

In the Indian arts, it is usual to find intricate works connected to the fractal geometry the most famous are: the *mandala* and the *kolam*. Mandala in Sanskrit means "circle." But, if you look more closely you find that the word Sanskrit "manda" means "essence" and the suffix "la" means "container" that explains the etymology of mandala, or "container of essence," circle or sphere containing the essence. It is believed that mandala is a place where enlightened beings dwell, and that from this place they give strength to the initiates. The mandala is the symbol of the universe and its energy, it is the macrocosm of the microcosm. It is the union and harmony of the self with the universe, the inner world and the outer world. This cosmic diagram is a support to meditation and initiation. There are four types of mandala: the three-dimensional mandala, the mandala painted, the concentration mandala and the sand mandala that is built by Buddhist monks. Mandalas' shapes remind to self-similarity properties (Sala, 2004; Sala and Cappellato, 2004). In Figure 7 three examples of mandalas.

The *kolam* is a decorative draw that embellishes the courtyards and the doorstep of the homes in the villages, temples and prayer rooms in South India. It consists of some small geometrical patterns repeated many times, that comprises lines, dots, squares, circles, triangles, lotus, shells, leaves, trees and flowers connected in very complicated ways. *Kolam* is an auspicious symbol, and it is the most important kind of Indian female artistic expression (Sala and Cappellato, 2004). The tradition of the *kolam* is handed down from mother to daughter, as an important ritual technique that is also appreciated as a demonstration of mental discipline and concentration skills. The forms that the *kolams* take are divided into groups, in relation to their use. There are kolams for a day of celebration, for ordinary days and for special recurrences.

Kolams are known by different names in different parts of India. *Muggulu* in Andrapradesh, *Hase* in Karnataka, *Chowkpurna* in Uttar Pradesh, *Rangoli* in Gujarat and Maharashtra, and *Alpana* in Bengal and Assam. Many ancient Sanskrit texts (about five millennia old) described the kolams, which can cover areas up to three meters by three meters.

Kolam patterns provide examples of cycle languages, and Rosenfeld (1975) studied a cycle grammar for generation and description of pictures with rotational symmetry.

Kolams can be defined using two-dimensional picture languages with formally definable syntactic rules. Siromoney et al., (1972) proposed the formal properties of these languages.

Siromoney et al., (1974) have also introduced different types of array grammars generating array languages, giving specific instructions for drawing certain kinds of kolam patterns using rewriting rules (Siromoney and Siromoney, 1987; Siromoney et al., 1989).

Figure 7. Mandalas' shapes remind the self-similarity.

Figure 8. *Kolams*: three examples (image in center, Source: By Nagata Shojiro/ InterVision Institute/KASF member - "nagata Shojiro's file", CC BY-SA 3.0, https://commons.wikimedia.org/w/index.php?curid=4204467).

Prusinkiewicz and Hanan (1989) have shown that many of the more elaborate kolams can be generated, as fractal objects, using the Lindenmayer systems, also known as L-systems (Prusinkiewicz and Hanan, 1989). Other studies on kolam patterns were done by Ascher (1991), Gerdes (1989), Nagata and Yanagisawa (2004, 2007), Robinson (2007).

In Figure 8 three examples of kolams; at right, there is a fractal kolam generated using L-systems the *Anklet of Krisna*; in center, there is the first application of a single cycle white line Kolam on cloth, the design of which shows nine goddesses of only symmetrical 3x3 single cycle Kolams, and a Swastik form. No gaps to be left anywhere between the line for evil spirits to enter. It is Hindu belief that the geometrical patterns and designs applied with rice flour at the entrance to a home, invites Goddess Lakshmi into the household, and drives away the evil spirits.

In classical Islamic Art, ornamentation has a significant value that can be seen in every artistic expression from the carved marble panels of grand Mughal doorways in India, to the blue ceramic tiles of *Masjids* in Iran, to the elegant decorative artefacts in Syria. Arabesque, its style, composition and principles can be found in every objet d'art of classical Islam. The characteristic of Islamic art is a preference for covering surfaces with patterns composed of geometric or vegetal elements like foliage, flowers, and an extensive use of abstract geometric designs (Blair and Bloom, 1995, 1997; Wilson, 1988; Stierlin, 2002; Bonner, 2003; Sala and Cappellato, 2004). One can find the principles of geometry along with a keen sense of balance in composition strongly embedded in Islamic art. El-Said and Parman present a system in which geometrical grids are broken down into identical units, which are repeated in regular sequence (El-Said and Parman, 1976). Figure 9 shows an Islamic dome with fractal decorations.

Jay Bonner analyzes the three traditional Islamic geometric ornaments, 14th and 15th century, where is present the self-similarity (Bonner, 2003). Bonner classifies three types of the self-similar Islamic geometric patterns. He classifies as: Self-Similar Type A Patterns, Self-Similar Type B Patterns, and Self-Similar Type C Patterns, respectively.

The Self-Similar Type A Patterns is characterized by a primary repetitive geometric pattern, with a reduced scale on a secondary geometric pattern, which has the same geometric characteristics as the primary, and it fills the complete background of the primary pattern. Figure 10a shows an example of Self-Similar Type A Patterns (Bonner, 2003, p.4).

The Self-Similar Type B Patterns, is realized on a primary geometric pattern, where the lines of which have been widened to a proportion that allows for a secondary geometric pattern, which has the same geometric characteristic of the primary pattern but at a reduced scale, to be placed within the widened lines (Figure 10b) (Bonner, 2003, p.3).

Self-Similar Type C Patterns is the third, and last type. It is present in Morocco and in Andalusia (Spain). The self-similar patterns of these regions are based on colour contrast to emphasize the primary design. An example of Self-Similar Type C Patterns is illustrated in Figure 10c (Bonner, 2003, p.4).

The artistic world of the 20th century had a short-lived art movement based on cubic and polyhedral forms that revolutionized the figurative language: Cubism (1907-1914). Two convergent influences can be found at the origin of Cubism: the lesson of the last Paul Cézanne (1839-1906) and the discovery of primitive arts (Iberian, African and Oceanic sculpture).

Figure 9. Dome with classical Islamic decoration (Dubai, United Arab Emirates, UAE), where the self-similarity is evident.

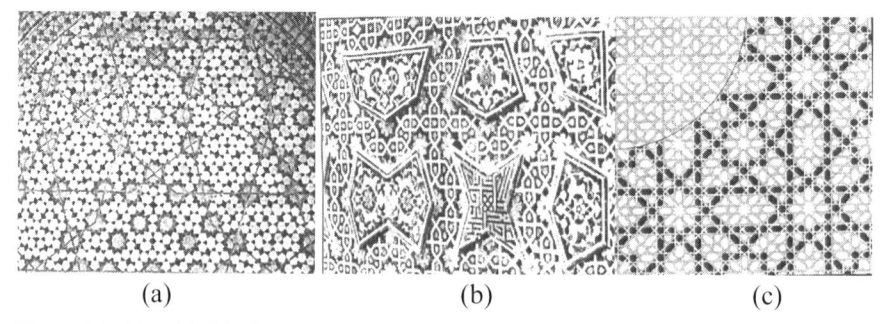

(a)　　　　　　　　　(b)　　　　　　　　　(c)

Figure 10. (a) Self -Similar Type A design from the Drab-i Imam (Isfahan, Iran). (b) Self -Similar Type B design from the *Masjid-i Jami* (Isfahan, Iran). (c) Self -Similar Type C design from the *Alcazar* (Seville, Spain) (Bonner, 2003, p. 4).

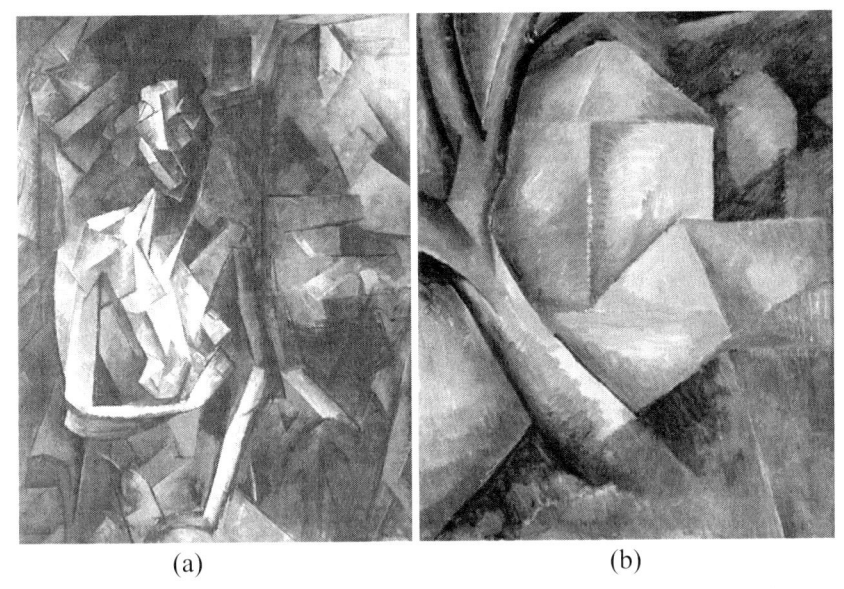

(a)   (b)

Figure 11. (a) *Figure dans un Fauteuil* (1909-1910, *Seated Nude*) (By Pablo Picasso - Tate Modern, London, PD-US, https://en.wikipedia.org/w/index.php? curid= 39252326). (b) *Maisons et arbre* (1908, *Houses at l'Estaque*), (By Georges Braque - Lille Métropole Museum of Modern, Contemporary and Outsider Art, PD-US, https://en.wikipedia.org/w/index.php?curid=44948498).

With reference to the pictorial research of Picasso (1881-1973) and Braque, cubism is usually divided into two main phases: a first defined "Analytic Cubism" and a second phase defined "Synthetic Cubism."

Analytic cubism is characterized by a process of numerous decompositions, and re-compositions, which gives to the painters of this period their unmistakable texture of differently crossed angles. Synthetic cubism, on the other hand, is characterized by a more direct and immediate representation of the reality that it wishes to evoke, completely canceling the relationship between figuration and space. Synthetic cubism, more than any other pictorial movement, revolutionizes the very concept of painting as being itself "reality" and not "representation of reality." Cubism consists of a split of multiple polyhedral forms and a reduction in colors to a range that includes gray, blue, beige and brown. In this decomposition of the forms, one could see a kind of fractality, as shown in Figure 11a and 11b, which reproduce respectively: *Figure dans un Fauteuil* (1909-1910, *Seated Nude*) by Picasso (Tate Modern, London), and *Maisons et arbre* (1908, *Houses at l'Estaque*), by Braque (Lille Métropole Museum of Modern, Contemporary and Outsider Art, Lille, France).

Maurits Cornelis Escher (1898-1972) is the Dutch artist who has most used math and geometry in his works. Some of his famous engravings can be explained using a "fractal" key, where the self-similarity is obvious (Sala and Cappellato, 2003a). For example, in his work entitled *Circle Limit IV*, the artist has made with reference to a Poincaré plane. However, if we analyze this Escher's work using the fractal geometry, we note that the angels and devils present in the engraving are self-similar, in fact decrease of scale, while maintaining their shape. Another example of fractality you can find it in irregular shapes, of Escher's studies related to the flooring of the *Cathedral of Ravello*, where Sierpinski triangles is present.

For the Spanish artist Salvador Dalì the self-similarity is used to emphasize the horror of war in his painting entitled *The Face of War* (1940-1941).

American painter Jackson Pollock (1912-1956), is the most emblematic representative of Action Painting, the current that represents the American contribution to the informal, which could be analyzed using a "chaotic and fractal" key. Taylor, Micolich and Jonas analyzed the Pollock's works using a fractal point of view (Taylor, Micolich and Jonas, 1999a). They also coined the term "fractal expressionism" to distinguish fractal art generated directly by artists from fractal art generated using mathematics or computers (Taylor, Micolich and Jonas, 1999b). Taylor have also analyzed the Mondrian's artistic production using the same fractal key (Taylor, 2007).

Klaus Ottmann is a museum curator. He organized an exhibition entitled *Strange Attractor: The Spectacle of Chaos*, in 1989. He believes that in contemporary art is going on a fractal revolution. He says (Briggs, 1992, p. 166):

> We might speak of a fractalist activity as we once spoke of a surrealist or a structuralist activity. Fractalist artists are both a mirror of the psychological and social state of society, and an interface. They no longer concern themselves with the mere manufacturing of objects but with experience of fractalization. Watch for the presence of any one of the three attributes of fractals (scaling, self-similarity, and randomness) to determine whether the fractalist vision is at work.

The artist Carlos Ginzburg, in his work entitled: *Fractal Chaos* (1985-1986), uses the self-similarity. He says (Briggs, 1992, p. 167):

> Understanding fractal and chaos made me more than change my perceptual experience of the world. My "Homo Sapiens," "Homo Faber," "Homo

Demens," "Homo Ludens" dimensions changed definitely into "Homo fractalus" one. I'm a fractal subject-fractalman.

The evolution of computer science and the new technologies have created new kind of artists. They use fractal algorithms to generate new shapes visualized on the computer's screen. Figure 12a shows an example of fractal art, it is generated by computer using fractal algorithms.

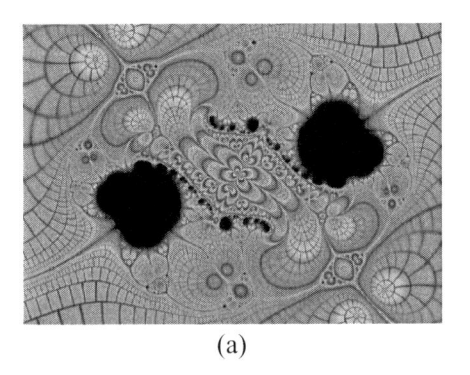

(a)

(b)

Figure 12. (a) An example of computer art based on fractals (b). Sculpture realized by John Edward (a clip is available at the following URL: https://vimeo.com/116582567).

Other artists use the computer, the fractal algorithms and the 3D printers (as output devices) to realize their works. For example, American artist and designer John Edmark. His artistic objects have an interesting property: they appear at the human eye as fluid objects and in ever changing. This suggestive effect is achieved by the correct use of lighting and rotation techniques. The sculptures, with self-similar shapes (an example in Figure 12b), are illuminated with stroboscopic lights and rotated on a plane, with a sequence following the golden rule, the same which we find in the growth of many natural elements (for example, shells, sunflowers, or pine cones).

The movement speed and the frequency of the flash are synchronized so that the beam of light hits the sculpture after each rotation of 137.5 degrees (the angular version of the golden section). Edmark called his sculptures: *Blooms*.

2. FRACTALITY IN ARCHITECTURE

In different cultures and in different styles are present many unconscious fractal components. An interesting example is the capital of an Egyptian temple column (Figures 13a and 13b). Ancient Egyptian cosmogony, often used to represent the development of the universe the white lotus flower. The lotus' corolla is organized in petals within petals within petals, in this way the lotus represented the cosmos on smaller and smaller scales. This is a clear example of self-similarity. We can compare the stylized lotus petals and the similarity of this representation to the first three steps of a Cantor's set (Figure 13c). Probably, it is the oldest man-made fractal. Eglash focuses his attention on the fact that African architecture reflects both the social structure and the religious structure of a settlement. The architectural examples presented by the author show that fractal features are a direct consequence of some structural or organizational features of a settlement. In some cases they are the representation of a religious or social hierarchy (Eglash, 1999). Eglash claims, in relation to a political perspective, that European urban planners consider African settlements as big villages and not real cities, as it does not use, as in Europe, Euclidean geometry for street tracing, and in urban organization, but rather complex shapes that resemble fractal geometry. He writes (1999, p. 196):

> Thus fractal architecture was used as colonial proof of primitivism …
> During the development of colonial cities, the chaos of African architecture was used as both symbol and symptom of European fears over social chaos.

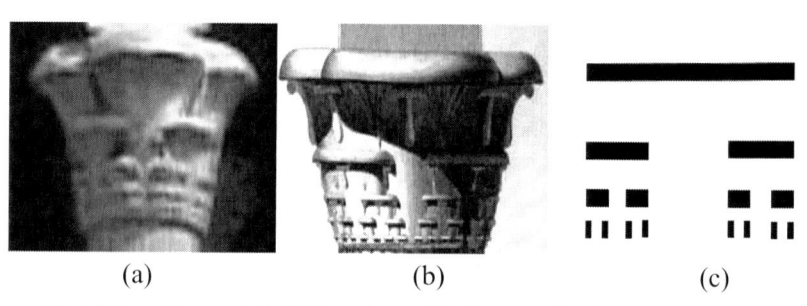

| (a) | (b) | (c) |

Figure 13. (a) Egyptian capital, (b) its scheme (c) shows and a Cantor set.

Eglash observed three interesting fractal organizations in the African settlements. They are circular fractals, rectangular fractals, and branching fractals. An example of circular fractals is provided by Ba-ila, which is located in Southern Zambia (Africa). Each house has a ring shape, with the cattle fence, and has both a main entrance and an exit. Near the main entrance, there are small buildings that serve as food dispensers. Each house is inside a fence, also with a ring shape, which delimits the settlement. Continuing through the ring the buildings, which keep all the same form, become larger, to arrive to the greater dimension which belongs to the father of this family community and which is located in an opposite position to the entrance main. In front of this house, there is the chief's house. Ba-Ila represents an example of self-similar settlement based on circular fractals. An example of rectangular fractals is in the city of Logone-Birni (Cameroon). The complex palace of the chief is organized following a subdivision based on rectangular fractals. Examples of branching fractals are in a Senegalese Settlement and in the city of Banyo (Cameroon).

Another example of African fractal architecture is the *Great Mosque of Djenné* (Djenné, Mali). It is an interesting example of the Sudano-Sahelian architectural style. The current structure of this mosque is dated in the first decade of 19th century (1907), but the first mosque on the site was built around the 13th century. It is built using adobo (or adobe). Adobe is a mixture of clay, sand, and often organic material, sun-dried straw used by many populations at all times to realize bricks. Figure 14 shows The *Great Mosque of Djenné*, where we can observe a fractality.

Hindu traditional architecture has more symbolic meanings than the architecture of other cultures (for example, European or American cultures). Hindu architecture is articulated. Quoting William Jackson (p. 14):

> The ideal form so gracefully artificed suggests the infinite rising levels of existence and consciousness, expanding sizes rising toward transcendence above, and at the same time housing the sacred deep within. The gated enclosures-within-enclosures enshrine the inner sanctum, which for Hindus holds an external likeness of the inmost depths of divine mystery.

Hindu philosophy considers the cosmos self-similar. Each fragment of the cosmos is believed to be whole in itself. A kind of fractal cosmos. The Hindu temples are designed and constructed as models of the cosmos, so they are fractal. Figure 15 shows a scheme, which resumes the fractal structures of the Hindu temples, where parts resemble the whole.

Figure 14. The *Great Mosque of Djenné* (Mali) (by Ruud Zwart - Photo taken by Ruud Zwart, CC BY-SA 3.0, https://commons.wikimedia.org/w/index.php?curid=3326719).

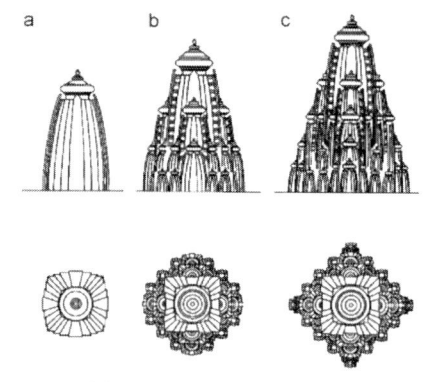

Figure 15. Fractal structures of the Hindu temples, where parts resemble the whole (by Ashish Nangia - http://www.boloji.com/index.cfm?md=Content&sd=Articles &ArticleID =900, CC BY-SA 4.0, https://commons.wikimedia.org/w/index.php? curid=52240374).

Figure 16. Temples of Prambanan. Of the three largest temples, one is dedicated to Shiva and it is in the center, Brahma on the left, and Vishnu on the right.

Another interesting example is the complex of 9th century Hindu temples *Prambanan* (or *Candi Prambanan* or *Candi Rara Jonggrang*).

This archeological site is located approximately 20 kilometres northeast of the city of Yogyakarta on the boundary between Central Java and Yogyakarta provinces (Indonesia). The complex of temples is dedicated to the Trimurti (the Trinity of the Hinduism). The Trimurti is the supreme expression of the divinity in the Hinduism, and it comprises the Creator: Brahma, the Preserver: Vishnu, and the Destroyer: Shiva. Figure 16 shows a view of the three temples, even in this case self-similarity is evident and repeated on several scales.

Figure 17 shows another example of Hindu fractal architecture: the *Mandara Giri Semeru Agung Temple* (*Pura Mandara Giri Semeru Agung*, Java, Indonesia). This recent Hindu temple was finished in 1992, and it is located at Senduro, Lumajang Regency, on slope of Gunung Semeru, the highest peak in Java.

Figure 18 shows The *Mother Temple* of Besakih, or Pura Besakih, in the village of Besakih on the slopes of Mount Agung in eastern Bali, Indonesia, is the most important, the largest and holiest temple of Hindu religion in Bali.

Figure 17. *Mandara Giri Semeru Agung Temple* (Senduro Village, Eat Java, Indonesia) has self-similar components.

Figure 18. *Mother Temple* of Besakih, or Pura Besakih, in the village of Besakih on the slopes of Mount Agung in eastern Bali, Indonesia (By Photo by CEphoto, Uwe Aranas, CC BY-SA 3.0, https://commons.wikimedia.org/w/index.php? curid= 38631708).

The *Buddhist Temple* of Borobudur (Java) was built on a hill in the 9th century under the Java Sailendra dynasty. It represents a mountain temple that has influenced the architecture of other temples, such as Angkor in Cambodia. The building consists of 10 terraces, one for each phase of the spiritual path to perfection. It is divided into three levels, which correspond to the three Buddhist spheres: the base represents life in the spirits of desire (kamadhatu), the five square terraces represent the progressive emancipation from the senses (rupadhatu), the three circular terraces represent the progression of the soul towards nirvana (arupadhatu). Everything is crowned by a large central stupa. The temple is conceived as a mandala, showing the universe in the Buddhist religion, the self-similarity is evident, as shown in Figure 19 (Sala and Cappellato, 2004). Figure 20 describes a cross section and ratio of Borobudur. The ratio containing the number 9, 6, and 4.

On the Asian Continent, Islamic architecture also used self-similarity, especially in arches and domes that repeat their shape on different scales.

The *Mausoleum* of Sitt Zubaida (1179-1225) in Baghdad, probably made by the Caliph an-Nassir on the tomb of his mother, has as its highlights a conical dome where overlapping niches are becoming smaller and smaller. This type of dome was common in hot and dry regions such as Lower Iraq.

Kãmadhãtu   Rüpadhãtu   Arüpadhãtu

Figure 19. The plan of Borobudur took form of a Mandala, a model of universe in Hindu-Buddhist cosmology. It consists of three ascending realms, Kãmadhãtu (the realm of desire), Rüpadhãtu (the realm of form), and Arüpadhãtu (the realm of formlessness). (By Gunawan Kartapranata - Own work, CC BY-SA 3.0, https://commons.wikimedia.org/w/index.php? curid=8763651).

Mughal architecture is an architectural style developed by the Mughals between the 16th, 17th and 18th centuries during ever-changing extent of their empire in Medieval India. This style shows influences by Persian, Islamic, Turkic and South-Asian architecture. Mughal buildings have large bulbous domes in different scales, slender minarets at the corners, large vaulted gateways and delicate ornamentation. We can find examples of this architectural style in India, Afghanistan, Bangladesh and Pakistan. Figure 21 shows *Humayun's Tomb* (1565-1572) in Delhi (India). Built for Humayun (1508-1556), second Mughal Emperor of Mughal Empire. The self-similarity is present, in this funeral monument, in the arches' shape and in the cupolas' shape repeated in different scales (Sala and Cappellato, 2003a).

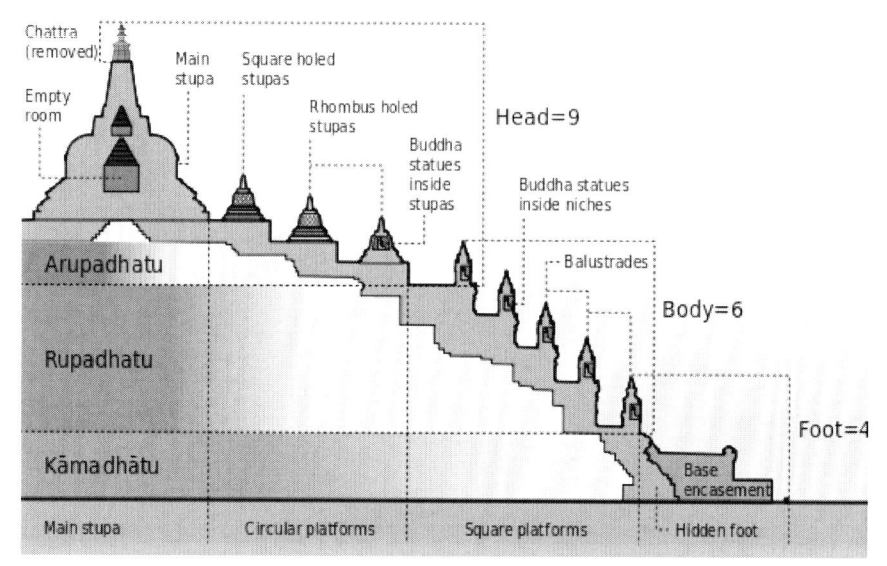

Figure 20. Cross section and ratio of Borobudur (By Gunawan Kartapranata - Own work, CC BY-SA 3.0, https://commons.wikimedia.org/w/index.php?curid= 17256197).

In Chinese and Japanese architectures, self-similarity can be found in the roofs of wooden pagodas (Sala, 2002; Sala and Cappellato, 2004). Their simple original structures consisted of a base on which a cusp was raised, which, with the passage of time, protrudes upwards to support a series of overhead roofs. Thus, the roofs multiply overlapping with a decreasing dimension, creating a prospective effect. The central plant of the pagodas usually has a polygonal shape.

Mesoamerican civilisations realized numerous administrative and ceremonial centres and erected numerous monuments (for example, pyramids and tombs), that reflected astronomic knowledge and expertise in numeration and calendars. In Mesoamerican architecture, we can find fractal components. The *Temple of Kukulcan* (c. 8th – 12th century), also known as *El Castillo*, is located in the center of the Chichen Itzá archaeological site (Mexico). It is a step-pyramid, and it shows fractal components (Figures 22a and 22b), which are also present in the *Temple of Great Jaguar* (*Temple I*) and in the *Temple II* of Tikal (Mexico) (Burkle-Elizondo et al., 2015).

Figure 21. *Humayun's Tomb* (1565-1572) in Delhi (India) shows fractal components.

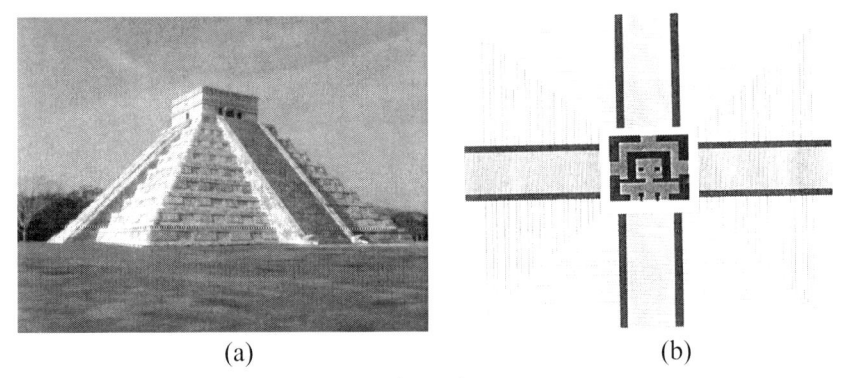

(a)	(b)

Figure 22. (a) *Temple of Kukulcan* (c. 8th – 12th century), also known as *El Castillo* (Chichen Itzá, Mexico). (b) The plan highlights fractal components.

In Europe, Gothic and Baroque architectures had complex decorations. Frequently, their cathedrals have self-similarity, which is expressed in several levels. The Gothic is a style developed in northern France that spread throughout Europe between the 12th and 16th centuries. "Gothic" was a term firstly used during the later Renaissance by the Florentine historiographer Giorgio Vasari (1511-1574) as in reference to the Nordic tribes that overran the Roman empire in the 6th century. He writes:

> Then arose new architects who after the manner of their barbarous nations erected buildings in that style which we call Gothic (of Gotthi).

Fulcanelli, the 20th century most enigmatic alchemist, gave another explication of the term Gothic, which is connected to the language of the alchemy (Fulcanelli, 2000).

Some fractal components are present in the Gothic churches; an example is shown in Figure 23, which reproduces the façade of the *Basilica di San Marco* (*St Mark's Basilica*, 1064-1093), Venice, Italy). The white arrows point out the fractal components (Sala and Cappellato, 2004, p. 139). The self-similarity is also present inside the Gothic Cathedral.

Figure 23. *Basilica di San Marco* (Venice, Italy) shows fractal components.

Figure 24. Fractal Venice: view of *Piazzetta San Marco* toward *Grand Canal* of Venice, at dawn, with *Doge's Palace* (9th-17th century) on the left and *Biblioteca Marciana* on the right (By Benh LIEU SONG - Own work, CC BY-SA 4.0, https://commons.wikimedia.org/w/index.php?curid=43315781).

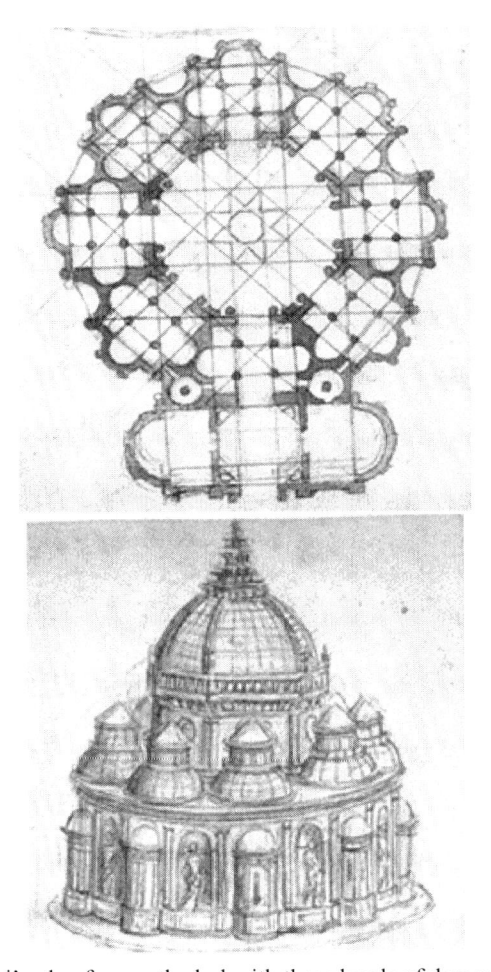

Figure 25. Da Vinci's plan for a cathedral with three levels of domes.

In the Italian Gothic style there are other examples which show the presence of the fractal components. In Venice there are many palaces (*Ca' Foscari, Ca' d'Oro, Palazzo Ducale (Doge's Palace)*, and *Giustinian Palace*) that have a rising fractal structure. For these features, Fivaz (1988) named this town: "fractal Venice," in Figure 24 (Fivaz, 1998; Sala and Cappellato, 2004).

Italian scientist Leonardo da Vinci (1452-1519) was a "Renaissance man," a polymath whose areas of interest included invention, painting, sculpting, architecture, science, music, mathematics, engineering, literature, anatomy, geology, astronomy, botany, writing, cartography, and history. He has made some very interesting projects of churches and cathedrals in six and nine domes, in which there is a clear repetition of the form of multiple scales. For this reason, we can think of fractality within these da Vinci's projects (Hersey, 1999; Sala and Cappellato, 2003). Figure 25 shows the da Vinci's plan for a domed cathedral with three levels of domes, where the self-similarity is present.

The Baroque (1600-1750) was born in Italy, and adopted in Germany, Netherlands, France, and Spain (Schneider, 2001). The term "Baroque" was probably derived from the Italian word "barocco," which was a word used by the philosophers during the Middle Ages to describe a hindrance in a schematic logic. After, this term was used to describe any contorted process of thought or complex idea. Another possible meaning derives by the Spanish "barrueco", Portuguese form "barroco," used to describe an imperfect or irregular shaped pearl. This word has survived in the jeweller's term "Baroque pearl." For other sources, the word "baroque" could be derived from the Arabic word of "burga", meaning "uneven surface" (Jairazbhoy 1965). Baroque was also associated with the Catholic art, but during the centuries it progressed and diffused its style into the Protestant countries. Baroque is a style that expressed power, and rigour, "the style of absolutism." Baroque favoured higher volumes, exaggerates decorations, and colossal sculptures (Stella, 1987; Hersey, 1999; Careri, 2003).

Baroque suggested movement in static works of art, and it influenced important challenges in architecture (Harbison, 2000). Baroque architecture was based on the mathematics (Hersey, 2000). In the age of the Baroque, the architects and the patrons thought of the buildings as "studies in practical mathematics" (this is a phrase of Italian religious Virgilio Spada (1596-1662), that has realized the plan of the *Chapel Spada* located in Rome) (Portoghesi, 1970; Magnuson, 1986; Hersey, 2000).

The Baroque architecture could be analysed using a fractal point of view (Sala and Cappellato, 2003a, Sala and Cappellato, 2003b; Bukdahl, 2017). For example, the self-similarity is present in the plans of some churches. For example, *Karlskirche (St. Charles's Church)* (1715-1737, Vienna, Austria) where the oval is repeated in three different scales (Sala and Cappellato, 2003b).

Swiss architect Francesco Borromini (1599-1667) used the octagons, the Greek crosses and other shapes for the coffering of the dome in his Roman Catholic church *San Carlo alle Quattro Fontane* (*Saint Charles at the Four Fountains*), also known as *San Carlino* (c. 1634-1644, Rome, Italy). Hershey (1999) analyses the lattice used to map a detail from Borromini's dome coffers in *San Carlo alle Fontane*. He describes the presence of two directional compressions, horizontal and vertical at the same time, over a (much shallower) dished plan. Borromini probably achieved the dome interior coffers using a particular technique. Figure 26 illustrates the *San Carlino*'s dome, where the ends of each lozenge and of each rhombus are unequal, the upper half of each octagon is smaller than the lower half, and the top of the upright in each Greek cross is shorter than the bottom of the lower part of the cross' upright (Hersey, 1999). These compressions introduces a type of self-similarity in the dome (Sala and Cappellato, 2003b).

We can find another example of self-similarity in the Baroque architecture in the cupola of the *Church of San Lorenzo* (1666-1680, Turin, Italy), designed by the Italian architect Guarino Guarini (1624-1683). About this church, Norwich writes (1975, p. 176):

> The Church of San Lorenzo, Turin, was begun by Guarino Guarini in 1668 for the Theatine Order, of which he was a member. The plan is remarkable for its curved bays pressing into the central domed space—an idea developed from Borromini—but the dome is even more remarkable. It is a masterpiece of ingenious construction—the ribs actually carry the lantern above them—which is also used to produce dramatic contrasts of light and shade.

Figure 26. The dome of *San Carlo alle Quattro Fontane* (Rome) by Borromini. It is a typical Roman Baroque architecture (by Welleschik – Own work, CC BY-SA 3.0, https://commons.wikimedia.org/w/index.php?curid=6843715).

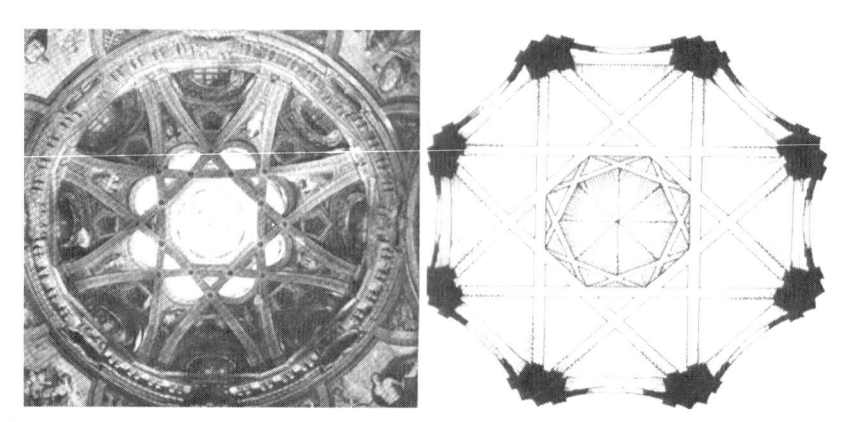

Figure 27. The dome of *San Lorenzo* (Turin, Italy) shows some self-similar components (Götze, 1988, p. 85).

Guarini used the octagonal star to define the bearing structure of the dome. The self-similar components are an octagon and an octagonal star, which are repeated in different scales, as shown in Figure 27 (Götze, 1988, p. 85).

Benoit Mandelbrot, father of fractal geometry, cites the *Paris Opera Building* (1865-1875) as an example of scaling symmetry that characterizes the fractal geometry. He observed, walking along the *Rue de l'Opera*, and approaching the building, the self-similar building details which reminds to the natural shapes. This comparison of Mandelbrot between the opera house and the natural objects emphasizes the intention of many artists to create forms that contain the vitality and the dynamism, which is typical in the natural forms (Briggs, 1992, p. 170).

Fractal geometry has inspired American architect Frank Lloyd Wright (1867-1959), one of the most important exponents of modern architecture. He made specific reference to the fractal geometry in the implementation of some of his projects, using a conscious fractality. For example, in the *Palmer House* (Ann Arbor, Michigan, 1950-1951) and in the *Marin County Civic Center* (San Rafael, 1957). In *Palmer House*, he used self-similar equilateral triangles in the organization of the project (Eaton, 1998). In *Marin County Civic Center* there is the presence of cycloid on four different scales of size. The building has an analogy of shape with a Roman aqueduct and bridge, with self-similar structures (Sala and Cappellato, 2004).

Danish architect Jørn Oberg Utzon (1918-2008) designed the *Opera House* (1957-1973) in Sydney (Australia) using unconscious fractal geometry through self-resembling forms that recall the valves of a shell, as shown in Figure 28. The *Opera House* was born from the assembly of three distinct parts: the stand, the auditorium and the covers (which had to be different portions of the same ideal sphere corresponding to a diameter of 75 meters).

Figure 28. *Sydney Opera House* (1957-1973) by Utzon. It is organized in self-similar way.

An unconscious use of fractal components are present in two works realized by Swiss architect Mario Botta. One is the *Library* (1985-1988) (Villeurbanne, France). Figure 29 illustrates the bottom view of the central light well. We can note the presence of self-similarity as the geometry of the well of light repeats its shape in the different floors, but with different dimensions. In this case, it is self-similarity repeated on four of different scales. On this project Mario Botta affirms (Sala and Cappellato, 2003a, p. 238):

> The feature of this building is the interior space formed by a concentric circular cavity that is spread to the entire height up to the cover. It represents the entire heart of the building as a button and a light column around which the volume recomposes its unity as a spiral wrapping in the definition of the different spaces. From the beginning, the idea was to create a central point. And this corresponds to the need I feel to find a "backbone" for each project, so that those who enter the building have the ability to orientate themselves. It can be a ray of light or a structural element.

The idea for this tender for the building was to create a house made of empty space within the solid building, with connecting staircases that are being wrapped around the vacuum.

This space was basically made up of an intuition of light.

Second "fractal work" is The *MART Museum of Modern & Contemporary Art* (1996-2002) in Rovereto (Italy). The volume of the MART is characterized by the large round square with the light filtering from above, a strong architectural sign, while the exhibition spaces are more anonymous. Botta comments on his choice, as follows:

> I think a good museum must answer these two contradictory aspects. In public space to have a great architectural strength to express its identity. In the exhibition galleries, let the works be talking.

The impressive mass of the museum is not revealed at first sight, but it remains partially hidden. The project follows an axial composition, and it is organized along the lane between two historical buildings. The lane leads to the circular plaza, covered by a glass dome, which recalls the self-similar shapes (Figure 30).

Figure 29. *Library* in Villeurbanne by Botta. Bottom view of the central well of light with the self-similarity.

A conscious use of fractal objects can be found in contemporary architecture, with the use of the computers, CAD/CAE systems, and new building materials. For example, Canadian-born American architect Frank Owen Gehry often uses fractal geometry, especially to highlight the rounded shapes and to create a plastic dynamism, as in his *Guggenheim Museum* of Bilbao (1991-1997), and in the *Walt Disney Concert Hall*, shown in Figure 31.

Figure 30. *MART* in Rovereto by Botta. Glass dome which highlights self-similarity.

Figure 31. *Walt Disney Concert Hall* (1999- 2003) Los Angeles, Frank O. Gehry.

Polish-American architect Daniel Libeskind has been fascinated by the fractal geometry and the self-similarity. In the *Art Museum* (Denver) and *Victoria & Albert Museum* (London) he uses them both to create light effects and to highlight the symmetry breaking. Rem Koolhaas, Zaha Hadid (1950-2016), Coop Himmelb(l)au, Jean Nouvel, Renzo Piano, Santiago Calatrava, Zvi Hecker, Steven Holl, Herzog & de Meuron, Massimiliano Fuksas, Tadao

Andō, Norman Foster and other "starchitects," find in the fractal geometry new shapes for their dynamic architectures.

Polish-born Israeli architect Zvi Hecker is a poet of form. In his works, he is recognized for an emphasis on geometry and modularity, but with an increasing asymmetry. Some of his major projects include the *Spiral Apartment House* (1984-1990) a Ramat-Gan (Israel), *Heinz-Galinski School* (1991-1995) in Berlin, the *Holocaust Memorial* (1997) in Berlin, the *City Center* (1996) in Bucharest, the Jewish Culture Center (1996-1999) in Duisburg (Germany). Fractal geometry is present in many of its projects as a stimulus and attempt to break symmetry, or to create effects that resemble self-similarity. The *Holocaust Memorial* is organized in self-similar structures, as if they were fragments of an explosion (Sala and Cappellato, 2004).

His project for the *Heinz-Galinski School* contains a triple fractality (Figure 32a). The first one is of a geometric type. We can observe fractal shape in analogy with the *Nude Descending a Staircase, No. 2*, by Marcel Duchamp (Figure 32b). The second is a fractality inspired by the nature. In fact, the school reproduces a sunflower and it represents a symbolic "gift," which the architect makes to the children of Berlin.

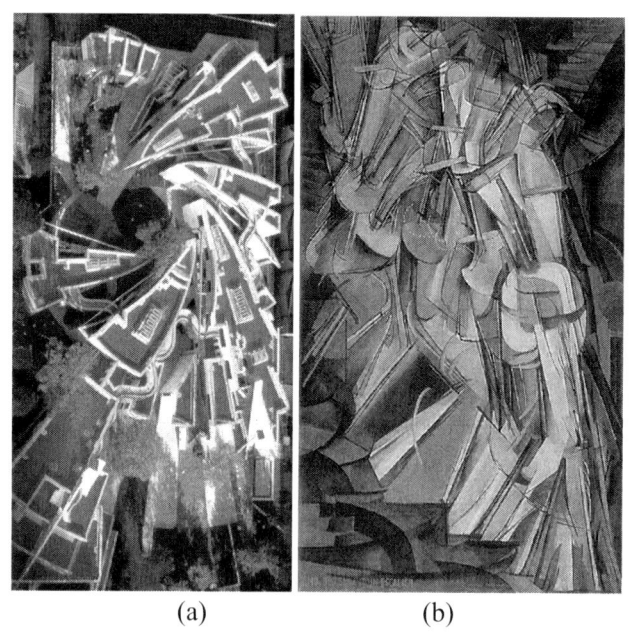

(a) (b)

Figure 32. (a) Heinz-Galinski School by Hecker. (b) Nude Descending a Staircase, No. 2, by Marcel Duchamp.

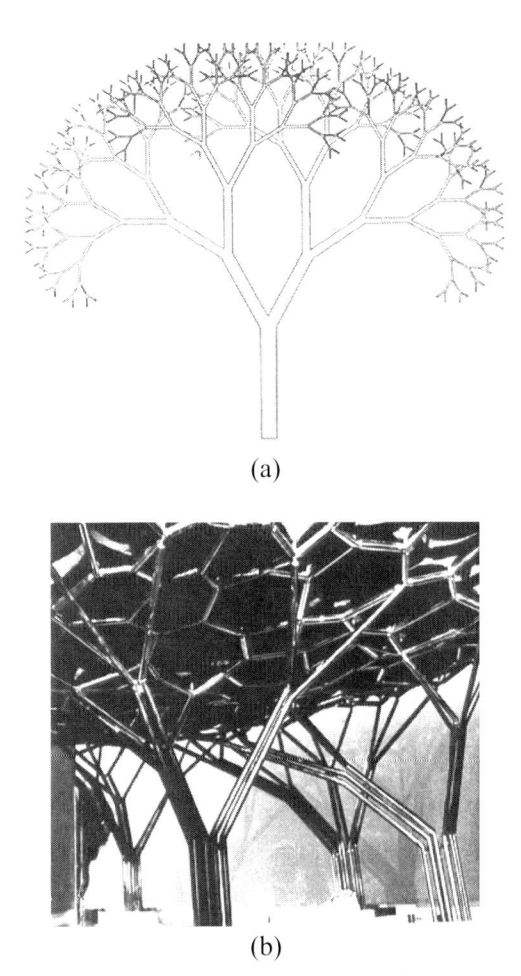

(a)

(b)

Figure 33. (a)A fractal tree. (b) Tensile structure designed by Otto Frei.

The third is a "logic" self-similarity: the school as a metaphor of a small city, in a model like a "Russian doll." In fact, Hecker affirms (Burke and Grosvenor 2008, p. 183):

> The school is a city within a city. Its streets meet at squares and the squares become courtyards. The walls of the schoolhouse also build walkways, passages, and cul de sacs. The outside of the school is also the inside of the city, because the school is the city.

Figure 34. *Gare de Oriente* (*Orient Station*) (Lisbon, Portugal). Often, Calatrava uses shapes that recall arboreal structures (By Bobo Boom - Gare do Oriente Uploaded by tm, CC BY 2.0, https://commons.wikimedia.org/w/index.php?curid =29907188).

(a)   (b)

Figure 35. (a) *Olympic Park Observation Tower* in Olympic green, Beijing (by Yu Lou - Own work, CC BY-SA 4.0,https://commons.wikimedia.org/w/index.php? curid=51196657). (b) A fractal tree.

Bifurcation processes, which are described through specific fractal systems defined as L-Systems, can be applied in architecture.

Figures 33a, and 33b respectively illustrate a fractal tree, and an example of tensile structure designed by German architect Otto Frei (1925-2015). Figures 34 reproduces the view of the Orient Railway Station of Lisbon, by Santiago Calatrava.

An interesting example of fractal tree is in contemporary Chinese architecture: the 244-meter-tall *Olympic Park Observation Tower* in Olympic green, Beijing (Figure 35a). Figure 35b shows another fractal tree.

CONCLUSION

This paper has analyzed the interconnections between fractal geometry, arts and architecture. In particular the property of the self-similarity and the processes of bifurcation. Interesting studies on the complex dynamics of visual arts, connected to the fractality, have been done by Kocić and Stefanovska (Kocić and Stefanovska, 2007). They claim (2007, p. 59):

> Art is the human response to the enigma of the Universe. The huge complexity of the Universe is reflected in the human minds. Some very "compressed" and fractal-like physical contents conserve selected information (mostly in pictorial form) and influence artists' creations. In this way, the products of art, during the mankind history, also resembles on a highly complex corpus that is characterized by the features of any other complex almost-fractal object, like presence of bifurcations, self-similarity of hierarchy of complexity.

Thus, fractal geometry can help to introduce new paradigms in the arts and in architecture (Fivaz, 1988; Briggs, 1992; Batty and Longley, 1994; Bovill, 1996; Frankhauser, 1994, 1997; Sala and Cappellato, 2003a, 2004; Sala, 2002, 2005, 2007; Harris, 2012, Williams and Ostwald, 2015a, 2015b). We have to consider that the fundamental studies are very complex, and they are connected to the theory of chaos and dynamic systems. The research of the building's self-similarity and of the processes of bifurcation in architecture are only two different approaches in the fractal small scale analysis applied in architecture.

Property of self-similarity can become a key to analyze the nature, the arts and architecture. Italian architect Paolo Portoghesi, on the self-similarity, affirms that in the history of architecture, the property of self-similarity, the "waterfall" structures, scaling, geometric fragmentation, are "instruments," compositional processes through which unity is pursued in the multiplicity and where it is often seen a Beauty Key (Portoghesi, 1999).

Fractal geometry and self-similarity can be thought as a code of the forms, also applicable to urban planning (Batty and Longley, 1994; Frankhauser, 1994, 1997). Historical analysis has shown that cities tend to behave in the same way in response to similar structural adaptation and structural demands, increasing through the progressive assembly of settlements that come from the point of view of spatial organization and relational, as copies of the global organism, in a kind of self-similarity.

REFERENCES

Ascher, M. (1991) *Ethnomathematics - A Multicultural View of Mathematical Ideas*, Pacific Grove: Brooks & Cole.

Batty, M., Longley, P. A. (1994). *Fractal Cities: A Geometry of Form and Function*, Academic Press, London and San Diego.

Blair, S. S. and Bloom, J. M., *The Art and Architecture of Islam 1250-1800* (Yale University Press, London, 1995).

Blair, S. S. and Bloom, J. M., *Islamic Art* (Phaidon Press, London, 1997).

Boito, C. (1860). *Architettura Cosmatesca (Cosmatesque Architecture)*, Tipografia Salvi e Compo, Milano.

Bonner, J. (2003). Three Traditions of Self-Similarity in Fourteenth and Fifteenth Century Islamic Geometric Ornament, *Isama-Bridge 2003 Conference Proceedings*, Granada, Spain, pp. 1- 12.

Bovill, C., *Fractal Geometry in Architecture and Design*, Birkhäuser, Boston, 1996.

Briggs, J., Fractals. *The Patterns of Chaos* (Thames & Hudson, London, 1992).

Bukdahl, E. M. (2017). *The Recurrent Actuality of the Baroque*, Controluce 2017, Denmark.

Burke, C., and Grosvenor, I. (2008). *School,* Reaktion Books, London, p. 183

Burkle-Elizondo, G., Sala, N., and R., David Valdez-Cepeda, Geometric and Complex Analyses of Maya Architecture: Some Examples. Williams K. and Ostwald M. J. (eds) *Architecture and Mathematics from Antiquity to the Future: Volume 1 Antiquity to the 1500s*, Birkhauser, 2015, pp. 113-125

Careri, G., *Baroques*, Princeton Univ Press, 2003.

Eaton, L. K. Fractal Geometry in the Late Work of Frank Llyod Wright: the Palmer House. Williams K. (ed.), *Nexus II: Architecture and Mathematics*, Edizioni Dell'Erba, pp. 23 - 38, Fucecchio (1998).

El-Said, I., and Parman A. (1976). *Geometric Concepts in Islamic Art*, Dale Seymour Publications, London.

Fivaz, R. (1988). *L'ordre et la volupté* [Order and voluptuousness]. Press Polytechniques Romandes, Lausanne.

Frankhauser, P. (1994) *La Fractalité des Structures Urbaines* [The Fractality of Urban Structures], Collection Villes, Anthropos, Paris.

Frankhauser, P. (1997) *L'approche fractale: un nouvel outil de réflexion dans l'analyse spatiale des agglomérations urbaines*, Université de Franche-Comté, Besançon [The fractal approach: a new tool of reflection in the spatial analysis of urban agglomerations, University of Franche-Comté, Besançon].

Fulcanelli, (2000). *Il mistero delle cattedrali e l'interpretazione esoterica dei simboli ermetici della Grande Opera* [The mystery of cathedrals and the esoteric interpretation of the Hermetic symbols of the Great Work], Edizioni Mediterranee, Roma.

Gerdes, P. (1989) Reconstruction and extension of lost symmetries: Examples from Tamil of South India, *Comp. Math. Applic.*, 17, No. 4–6, 791–813.

Götze, H. (1988). *Castel del Monte*, Hoepli, Milano.

Harbison, R., *Reflections on Baroque*, Reaktion Book, London, 2000.

Harris, H. (2012). *Fractal Architecture: Organic Design Philosophy in Theory and Practice*. University of New Mexico Press.

Hersey, G. (1999). *The monumental impulse*, The Mit Press, London.

Hersey G. (2000). *Architecture and Geometry in the Age of the Baroque*, University of Chicago Press, Chicago.

Jackson, W. *Hindu Temple Fractal*, retrieved, 6/10/2017 from: https://www.academia.edu/347639/Hindu_Temple_Fractals.

Kocić, L. and Stefanovska, L. (2007). Complex Dynamics of Visual Arts. Sala N. (ed.). *Chaos and Complexity in the Arts and in Architecture*, Nova Science Publishers, New York, pp. 39-61.

Magnuson T. (1986). *Rome in the Age of Bernini*, Almquist and Wickell, Stockholm.

Nagata, S., and Yanagisawa, K. (2004) Attractiveness of Kolam design— Characteristics of single stroke cycle, *Bulletin of the Society for Science on Form in Japan*, 19, No. 2, 221–222.

Norwich, J. J. (1975) (ed.). *Great Architecture of the World*. Mitchell Beazley Publishers, London.

Portoghesi, P. (1970). *Roma Barocca: the History of an Architectonic Culture*, The MIT Press.

Portoghesi, P. (1999), *Nature and Architecture*, Skira, Milan.

Prusinkiewicz, P., and Hanan J. (1989). *Lindenmayer Systems, Fractals, and Plants* (Springer-Verlag, New York, 1989).

Robinson, T. (2007). Extended Pasting Scheme for Kolam Pattern Generation. *Forma Journal,* 22, pp. 55-64.

Rosenfeld, A. (1975), A note on cycle grammars, *Inform. Contr.*, 27, 374–377.

Sala, N. (2002). The presence of the self-similarity in architecture: some examples, M. M. Novak (ed.), *Emergent Nature*, World Scientific, 2002, pp. 273-283.

Sala, N. (2004). *Fractal Geometry in the Arts: An Overview across the Different Cultures.* Novak M. M. (ed.) Thinking In Thinking In Patterns Fractals and Related Phenomena In *Nature*, World Scientific, Singapore, pp. 177-188.

Sala, N. (2007) (ed.) *Chaos and Complexity in the Arts and in Architecture*, Nova Science Publisher, New York.

Sala, N., and Cappellato G. (2003b) The generative approach of Botta's San Carlino. *Proceedings 6ᵗʰ Generative Art Conference*, Milano, Italy, pp. 328-337.

Sala, N., and Cappellato G., (2003a). *Viaggio matematico nell'arte e nell'architettura* [Mathematical journey in art and architecture], Franco Angeli, Milano.

Sala, N., and Cappellato, G. (2004). Architetture della complessità. La geometria frattale tra arte, architettura e territorio [Architectures of complexity. Fractal geometry between art, architecture and territory], Franco Angeli, Milano.

Siromoney, G., and Siromoney, R. (1987) Rosenfeld's cycle grammars and kolam, in *Lecture Notes in Computer Science* 291, pp. 564-579, Springer.

Siromoney, G., Siromoney, R., and Krithivasan, K. (1972) Abstract families of matrices and picture languages, *Computer Graphics and Image Processing*, 1, 234–307.

Siromoney, G., Siromoney, R., and Krithivasan, K. (1974) Array grammars and kolam, *Computer Graphics and Image Processing*, 3, 63-82.

Siromoney, G., Siromoney, R., and Robinson, T. (1989) Kambi kolam and cycle grammars, in A Perspective in Theoretical Computer Science (Ed. R. Narasimhan), *Series in Computer Science,* Vol. 16, pp. 267-300, World Scientific.

Stella, J. (1987). *Baroque Ornament and Designs*, Dover Publications.

Stierlin H. (2002). *Islamic Art and Architecture: From Isfahan to the Taj Mahal, Thames & Hudson*, New York.

Taylor, R. P. (2007). Pollock, Mondrian and Nature: Recent Scientific Investigations, Sala N. (ed.). *Chaos and Complexity in the Arts and in Architecture* Vol. 1, pp. 25-37.

Taylor, R. P., Micolich, A. P. and Jonas, D. (1999a). Fractal Analysis of Pollock's Drip Paintings. in *Nature*, Vol. 399, p. 422; June 3, 1999.

Taylor, R. P., Micolich, A. P., and Jonas, D. (1999b). Fractal Expressionism, *Physics World*, 25, October 1999.

Williams, K., and Ostwald M. J. (eds) (2015a). *Architecture and Mathematics from Antiquity to the Future: Volume 1, Antiquity to the 1500s,* Birkhauser,

Williams, K., and Ostwald M. J. (eds) (2015b). *Architecture and Mathematics from Antiquity to the Future: Volume 2, The 1500s to the Future,* Birkhauser,

Wilson, E. (1988). *Islamic Designs for Artists and Craftpeople* (Dover Publications, New York, 1988).

Yanagisawa, K. and Nagata, S. (2007). Fundamental Study on Design System of Kolam Pattern. *Forma Journal*, 22, pp. 55-64.

Chapter 2

ALGEBRAIC CURVES AND VARIETIES IN 20TH CENTURY ART

Vincenzo Iorfida[,1], Mauro Francaviglia[†,2] and Marcella Giulia Lorenzi[‡,3]*

[1]School "Archimedes", University of Calabria, Campus of Arcavacata, via P. Bucci, Rende CS, Italy
[2]Dipartimento di Matematica, Università di Torino, via Carlo Alberto, Torino, Italy
[3]CEL – Lab. for Scientific Communication, University of Calabria, Campus of Arcavacata CEL. Ed. Polifunzionale, Rende CS, Italy

ABSTRACT

Algebraic curves and varieties have acted as sources of inspiration for artistic themes in many of the "geometrical forms" of Modern Art, in the XIX and XX Centuries. Nowadays, their beautiful shapes can be easily constructed by computers. We shortly discuss their role, providing a few examples from Painting to Sculpture and Architecture.

Among the most important conceptual structures developed by mankind, Geometry has always occupied a fundamental position of the greatest

---

[*] Corresponding Author: Vincenzo Iorfida, PhD, Email: unilmat@gmail.com.
[†] Email: mauro.francaviglia@unito.it († Prof. Mauro Francaviglia died 24 June 2013 RIP).
[‡] E-mail: marcella.lorenzi@unical.it.

importance; scientists, artists and philosophers saw in the rigor of geometrical methods the triumph of the artistic beauty and of the human thought.

In the 19th Century the study of algebraic curves leads to a new concept of "*space environment*", free and independent of the concepts of Euclidean Geometry, as a reference for the classification of algebraic varieties in a stricter way. A (real) algebraic curve is a surface in 3-dimensional (or possibly also higher D-dimensional) real space R^3, that and it can be defined as the set of zeros of a "*polynomial equation of degree n*". When the curve does not present "*singularities*" it can be seen as smooth surface (the differential of its equation must be non-zero at each point). However from an artistic viewpoint singular algebraic curves (and surfaces) are more "*interesting*" in the immediate vicinity of their singular points, where intricate behaviors appear, appealing to the eye-cusps, self intersections, nodes, and so on (Iorfida, Francaviglia, & Lorenzi, 2011a).

Along with the attention given to them by many mathematicians of the XX Century - that was at first mainly focused on the study of cubic and quartic curves - these figures captivated also the Art world.

As an example the of Clebsch's surface (Figure 1) is a smooth cubic that can be described by five equations of planes through the following:

$$x^3+y^3+z^3+(x+y+z)^3-(2x+2y+2z)^3=0.$$

In Physics, instead, quartic curves were the subject of special studies, especially in view of Optics and, more recently, of applications to other physical models (e.g., so-called String Theory). Both Kummer E. E. and Klein F. studied their properties, until reaching the most fascinating conclusions on properties of cubics. Famous was the "*Klein quartic*" and also the "*Kummer surface*" (a conical surface with 16 singular points; see Figure 2). Even mathematicians had to resort to "modeling" in order to remedy to the difficulties encountered in intuitive representations of these curves in the plane (Hartshorne, 1997).

At beginning of the 20th Century Tatlin V. E. and Rodčenko A. M., in former Russia, founded the so-called Constructivism, a movement of avant-garde of Art and Culture, that characterized the artistic world of the moment revisiting the relationship between space and time through "*dynamic searches*".

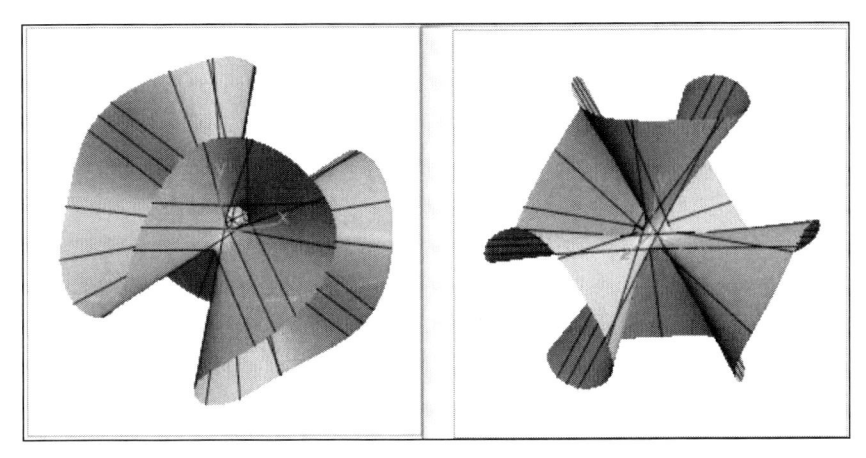

Figure 1. The Clebsch's Cubic - by Iorfida V.

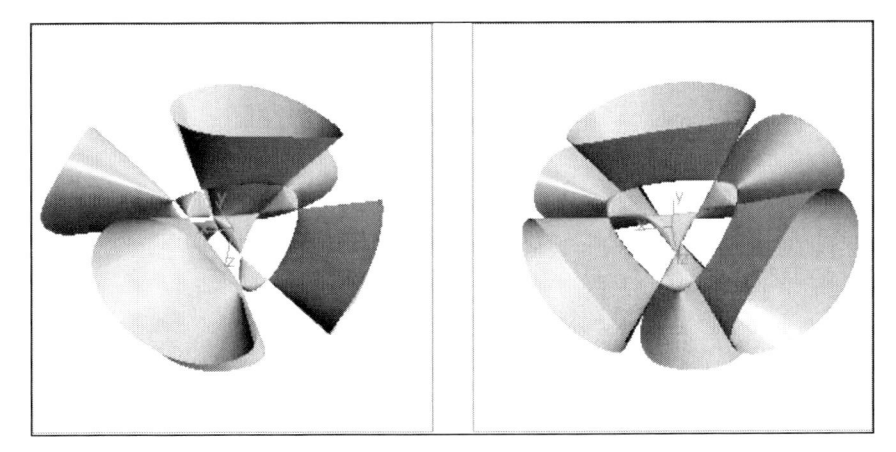

Figure 2. The Kummer's Cubic - by Iorfida V.

Constructivists, attracted by Sciences and Mathematics and inspired by architectural structures, reached new "accurate" forms depicting articulated models with complex shape, as e.g., it stands out from the project of the monument to the third International (never realized) by Tatlin V. E., consisting of two large metal spirals revolving in opposite directions and evolving to circumscribe a conical volume. Each geometric shape becomes an innovative model of artistic "retro" inspiration, as announced in 1923 by Tarabukin N. in the book "From the easel to the machine". The first sculptures of Gabo, were influenced by Cubism (Vierling-Claasen, 2010), as for example in the "Head No 2" of 1916 (Phillips Collection, Washington): it recalls surface models made of assembled material, composed of sheets of cut metal, cardboard and celluloid. The "Study for a Stone Carving" of 1933 resembles the model of a grooved surface and suggests that Gabo, being a careful researcher of new Art forms, could have visited and taken inspiration by the Poincaré Institute during his stay in Paris. The sculptor Pevsner A., Gabo's brother - even if he denied any direct mathematical influence on his work - with the series *"Developable Surface"* of the mid-30's refers to specific models of ruled surfaces. In fact, these works seem to be built just as the development of sets of lines. Even in France with A. Breton flourished an artistic and cultural current which was intended to be based on spontaneous and random elements, somewhat related also to the *"metaphysical painting"*. On the occasion of the exhibition of surrealist objects at the Galerie Charles Retton in Paris in 1936, Breton wrote an essay on *"The Crisis of the Object"*. Breton affirms: *"The laboratories of mathematical institutions around the World already show alongside with each other objects built according either to Euclidean or to non-Euclidean principles: both apparently bewilder the layman, but nevertheless they possess a fascinating and ambiguous relation to one the other in Space as we generally conceive it"*(Breton 1972).

Also the Surrealists, while not admittedly, made use of algebraic surfaces as models, since they are *"generated"* by algorithms and, then, they are *"scientifically rational"*. Ray Man and Max Ernst, both prone to Science and Mathematics, spread one impressionistic style in art and photographic movies. In his mathematical objects Ray Man proposed in fact surface models inspired by photographs that appeared in the 1936 edition of Cahiers d'Art, accompanied by an essay by Zervos C. (a Greek Art reviewer) concerning the interactions between Mathematics and abstract Art (Fortuné, 1999). At the International Surrealist Exhibition held in London in those years, the catalog explicitly highlights the appeal of mathematical models on the movement: the cover reproduces in fact a collage of surface models created by Max Ernst (a *"reptile-headed statue with models of algebraic surfaces in its hands"*), including the "Kuen surface"; as well as several other collages and paintings including the *"Feast of the Gods"* and *"The Chemical Wedding"* (both of 1948), *"Youngman Intrigued by the Flight of a non-Euclidean Fly"* (1942-47), which contain forms that recall the surface as a *"fuse ciclid"* and a *"ciclid horn"*. This, still referring to geometric techniques, to allude to a mathematical description of the designed World: In his artworks (as *"Design in Nature"*, *"The phases of the Night"*) through his *"dripping paint"* he always refers to an *"experimental Mathematics"*. In this direction we should also mention the notable artworks of Dalí Salvador Domènec Felip Jacint, who was also interested in Mathematics as clearly shown by some of his paintings (a famous collection of 1950) where the subjects are represented as being composed by rhinoceros horns (a form that fascinates Dalí, as its geometry evolves according to a logarithmic spiral). Or as in the celebrated *"Corpus Hypercubus"* (1954) where one can observe the development of a 4-dimensional (hyper)cube in the three-dimensional Euclidean space (Fig. 3), a solid that entails the fourth dimension. Hinton C. H., inventor of several neologisms (*"tesseract"*, *"kata"*...), was the first to catch a glimpse of the fourth dimension to describe four-dimensional directions. In his representation of a four-dimensional cube (as it can be shown in three dimensions) he started from an ordinary cube in three dimensions and consider its development in two dimensions by opening up its 6 faces (square) on the floor. In a completely similar way, *Dali* opens up a 4-dimensional cube by presenting its 3-dimensional realizations by 3-dimensional cubes glued along common (surface) edges. In this surrealistic and evoking painting he follows, in a suitable sense (Lorenzi & Francaviglia, 2011) also the path traced by the findings of Einstein A. on four-dimensional SpaceTime, but also the description of Plato's dialogue *"Timaeus"*, where it established that the five *"Platonic solids"* together with Empedocles' four elements (Fire, Air, Water and Earth) together with the *"quintessence"* form the structure of the Universe (Sala & Cappellato, 2003).

Also the paintings of Kandinskij V. and the Möbius shaped sculptures of Bill M. (Bill, 1949; Bill, 1949) do in fact represent lines and surfaces as algebraic varieties in two or more dimensions. Some of the physical phenomena that are difficult to be understood in three dimensions may find a simpler explanation, and even become trivial, if we imagine them to be in four or more dimensions.

For example, in the four-dimensional space of Kaluza T. (Schrödinger, 1950) for a review on, where the laws of Gravity and of Maxwell's electromagnetism find an interesting though physically inappropriate attempt of unification; and in the 5-dimensional *"hyperspace"* of Klein O., where extra-dimensions are seen as curves being curled up. Fresh ideas that, even in spite of their mathematical inability to generate a real *"unification"*, would lead in the XX Century to the unification intrinsic of *"Gauge Theories"* and, eventually, to the beauty of *"Superstring Theory"* as we know it today (Greene, 2000).

Modern techniques for the mathematical representation of multi-dimensional spaces and objects *"parallel coordinates"* introduced by Inselberg A. in the '80s, through which an hypersphere, passing into three dimensional space, would be perceived as a ball that comes out of nowhere, expands and then decreases until it disappears. Computer Art has allowed to represent visually this (Banchoff, 1990).

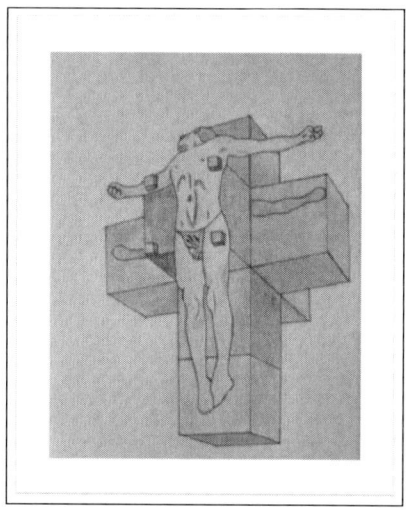

Figure 3. The Hypercube: the Grande Arche in Paris and Tesseract (photo© and picture© by Iorfida V.).

To date, advanced mathematical algorithms make it possible to study from the scientific point of view, generate and visualize by computer graphics the polytopes in four or higher dimensions present in both the physical and mathematical elements of architectural and artistic objects (Iorfida, Francaviglia, & Lorenzi, 2012).

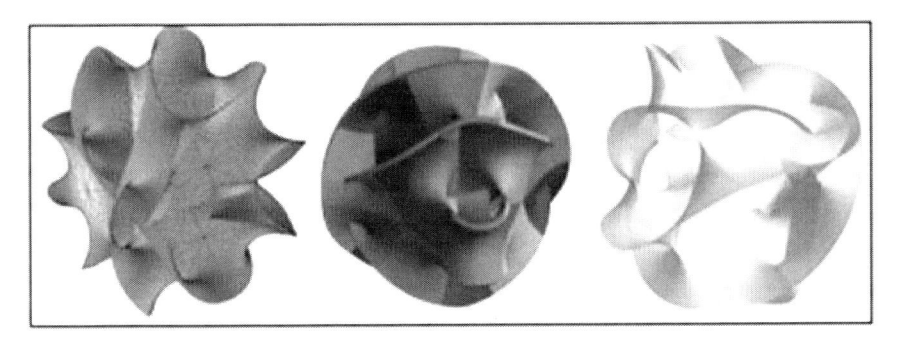

Figure 4. Calabi-Yau. The first and the third surface have been downloaded from the Wikimedia Commons. The authors are respectively Floriang and Jbourjai. The second surface is part of a work of Wolfram Research, Inc., by Jeff Bryant and A.J. Hanson. This file is licensed under the Creative Commons Attribution-Share Alike 2.5 Generic license.

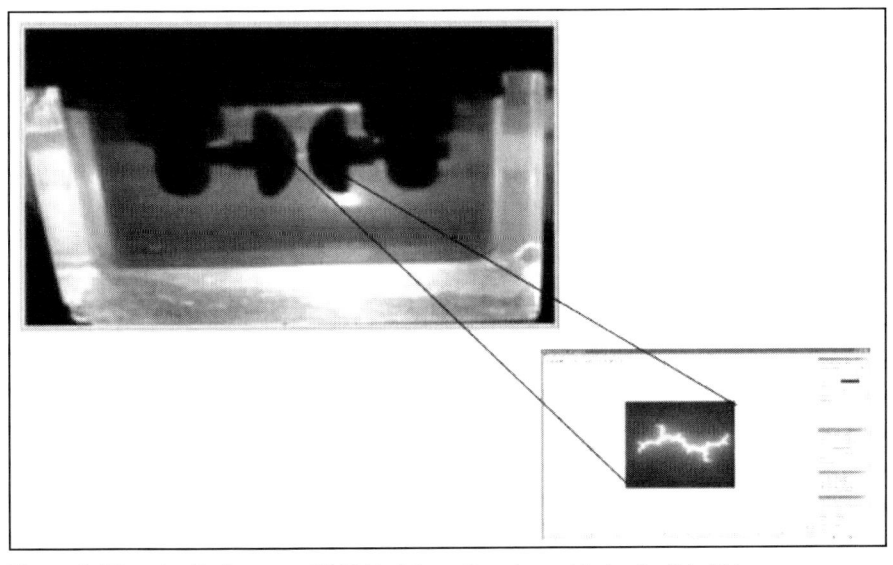

Figure 5. Electric discharge at 80 kVA (photo© and graphic by Iorfida V.).

CONCLUSION

In this framework it is, in fact, possible to approach, from a "tabular" and graphic viewpoint, in a rigorous way and of the base on modern techniques (Iorfida, Francaviglia, & Lorenzi, 2011b), the study of algebraic varieties in their "dynamical" and "evolutive" form; as an example, Calabi-Yau spaces embedded into 10 dimensions by String Theory, keeping also into account that already in R^3 (Fig. 4) there are over 500×10^6 of such varieties (Bini & Favale, 2012).

It is equally fascinating to know that such digital methods allow the possibility of finding new examples of algebraic varieties in higher dimensions, to understand their geometric properties up to create a library to classify them according to precise mathematical constructions. Discovering hidden extra dimensions induce us to obtain well defined and spectacular artistic shapes, only vaguely imaginable by making reference to those already known and represented. Even fractal ones…: passing from "standard spaces" to fractals is, in fact, almost immediate (Sala & Cappellato, 2004).

Therefore we have drawn our attention to these aspects too, with the aid of laboratory activities; this is related with new fractal artistic forms generated by physical phenomena as, e.g., "short circuits" (Fig. 5) - (Iorfida, Lorenzi, & Francaviglia, 2013), within which we have been able to observe "short-circuital forms" representing complicated surfaces at different magnifications, such as having been generated by a precise algorithmic-mathematical model able to create new and fascinating artistic shapes.

REFERENCES

Banchoff, T. (1990). *Beyond the Third Dimension.* New York, USA: Scientific American Library, W.H. Freeman & Co.

Bill, M. (1949). The Mathematical Approach in Contemporary Art. *Werk* (3), 74.

Bini, G., & Favale, F. (2012). Groups Acting Freely on Calabi-Yau Threefolds Embedded in Products of del Pezzo Surfaces. *Advances in Theoretical and Mathematical Physics, 16*, 887-933.

Breton, A. (1972). *Surrealism and painting.* London: Macdonald and Co.

Fortuné, I. (1999). Man Ray et les Objets Mathématiques. *Études photographiques* [Photographic studies] (6), 1-15.

Greene, B. (2000). *The Elegant Universe: Superstrings, Hidden Dimension, and the Quest for the Ultimate Teor.* New York: WW Norton & Company.

Hartshorne, R. (1997). *Algebraic Geometry* (1st ed.). Heidelberg, Germany: Springer-Verlag.

Iorfida, V., Francaviglia, M., & Lorenzi, M. G. (2011a). Algebraic Varieties in the of nineteenth century: from the concept of hiperspace to "exact" rational art. *APLIMAT – Journal of Applied Mathematics, 4* (4), 197-204.

Iorfida, V., Francaviglia, M., & Lorenzi, M. G. (2011b). Fractal Aesthetics in Geometrical Art Forms. *Proceedings of Bridges: Mathematics, Music, Art, Architecture, Culture* (p. 467-470). Phoenix, AZ, USA: Tessellation Publishing, Sarhangi R. and Sequin C.

Iorfida, V., Francaviglia, M., & Lorenzi, M. G. (2012). Communicating through the geometry of polytopes. *APLIMAT - Journal of Applied Mathematics, 5* (1), 85-90.

Iorfida, V., Lorenzi, M. G., & Francaviglia, M. (2013). The Probabilistic Behavior Generated in a Short Circuit by Elementary Particles. *Proceedings APLIMAT 2013*, 34-42.

Lorenzi, M. G., & Francaviglia, M. (2011). Geometry as a Source of Inspiration in Contemporary Art. *Bridges Coimbra Proceedings* (p. 365-372). Phoenix, AZ USA: Tessellation Publishing, Sarhangi R.

Sala, N., & Cappellato, G. (2003). *Viaggio nella matematica e nell'architettura* [Travel in mathematics and architecture]. Milano, Italia: Franco Angeli.

Sala, N., & Cappellato, G. (2004). *Architettura della Complessità* [Architecture of Complexity]. Milano, Italy: Franco Angeli.

Schrödinger, E. (1950). *Space-Time Structure.* Cambridge, UK: Cambridge University Press.

Vierling-Claasen, A. (2010). Models of Surfaces and Abstract Art in the Early 20th Century. *Proceedings Bridges 2010: Mathematics, Music, Art, Architecture, Culture* (p. 11-18). Phoenix, AZ, USA: Tessellations Publishing, Hart George W. and Sarhangi R.

In: Chaos and Complexity in the Arts …
Editors: N. Sala and G. Cappellato

ISBN: 978-1-53612-995-3
© 2018 Nova Science Publishers, Inc.

Chapter 3

FROM QUANTUM MODELS TO MATHEMATICAL AND CONCEPTUAL STRUCTURE OF THE 20TH CENTURY ART

Vincenzo Iorfida[*]

University of Calabria, Campus of Arcavacata, Via P. Bucci,
Rende CS, Italy

ABSTRACT

The 20th century represents a moment of total change for Science, as Renaissance for art. Keplero, Galileo and Newton's inviolable doctrines in Classical Physics are discussed. The Theory of Relativity presents space-time as an indissoluble structure, while Quantum Mechanics presents an undetectable universe, the subatomic one.

Science and its new counterintuitive models are synchronized with a creative art form dedicated to spread the difference between actual reality and the detected one. Both realities propose again the ancient dilemma of the dichotomy "continuous - discrete". In this work, through quantum nanotechnologies evolution, we approach an art world devoted to the entanglement phenomenon modifying the perceptual conception of space, and offering new horizons in the knowledge transfer in educational and divulgation fields.

---

[*] E-mail address: unilmat@gmail.com.

INTRODUCTION (SARDANASHVILY, 1995)

Since the 19th century, Physics has been supported by a rigorous and elegant mathematical structure, like that of Hamiltonian formalism [1]. Among the fathers of Quantum Mechanics, Max Planck occupied the main role. At the beginning of the 20th century, after some researches carried out on black bodies' radiations (2), it was deduced that energy was understandable as inseparable sets. These sets as discrete and inseparable unities were named "quantum". Therefore, on the 14th of December 1900, Planck presented a memory called "ZurTheorie des Gesetzes des Energieverteilung in Normalspektrum" to the Berlin Science Academy. Moreover, studies on the electromagnetic distribution of energy led him to the discovery of a universal physical constant, represented with the h symbol (called Planck constant), present in the quantum formula of energy E, contained in a frequency radiation f.

$E=hf.(^1)$

This constant was used by Albert Einstein to study the photoelectric effect and firstly appeared in a work published on the 18th of March, 1905 ("Übereinen die Erzeugung und Verwandlung des Lichtes betreffen den heuristischen Gesichtspunkt"). It opened new questions leading to the study of the infinitely small and to the Quantum Theory. The science aims were not directed to the macro-world described by Plato and Euclid, but to an unexplored world through new perspectives, e.g., the subatomic one and its laws. In 1913, Niels Henrik David Bohr took into account Planck's discoveries and, on the basis of three postulates, proposed a model of atom explaining both the substance stability and the emission spectrum of hydrogen atom, offering a crucial contribution to the development of Quantum Mechanics [2]. Afterwards, in 1924, the French physicist Luis-Victor Pierre De Broglie introduced the dualism of wave-particle theory in there search of elements inducing a natural phenomenon's symmetry.

However, according to Copenhagen, probabilistic models conquered an essential role in the quantum physic theory. After the study in depth of Bohr's ideas, Werner Karl Heisenberg was the first to formalize the quantum mechanics from a mathematical point of view, publishing "Überber die quantentheoretische Umdeutungkinematischer und mechanischer Beziehungen" on the 29th of July, 1925. Hence, Born realized that the model described by Heisenberg used a non-

---

(^{1}) The constant h is equal to $(6,62606957\ (29)*10^{-34}\ J*s.)$.

commutative algebra of infinite matrices, as well as the theoretical dissertation was not able to visualize the described phenomena. Therefore, the first complete formulation of Quantum Mechanics was represented in the work "Quantenmechanik II", published on the 16th of November, 1925, by Heisenberg, Born, and Jordan. On the 23th March, 1927, Heisenberg published "Über den anschaulichen Inhalt der quantentheoretischen Kinematik und Mechanik". In this work, he formulated the uncertainty principle: *"At the moment of the position determination, that is when the quantum of light is being diffracted by the electron, the latter changes its momentum discontinuously. This change is greater the smaller the wave length of light, that is the more precise the position determination. Hence, at the moment when the position of the electron is being ascertained, its momentum can be known only up to a magnitude that corresponds to the discontinuous change; thus, the more accurate the position determination, the less accurate the momentum determination and vice versa".*

According to the axiom, pairs of associated observable size (i.e., the motion quantity, and the position of a particle) were not measurable at the same time with absolute precision, because the absolute measure of the first produces uncertainty in the measure of the other one. In fact, in a simple system (composed by a single particle) the product of the uncertainties of the two measures can't be minor of h (Planck constant). In autumn 1927, during the congress of Como (Italy) for the memorial of Alessandro Volta's death, Bohr and Heisenberg formulated the "interpretation of the school of Copenhagen", and the complementarity principle and the indetermination principle became the pillars of the "official" physic interpretation of quantum mechanics [3].

Heisenberg (as Bohr) affirmed the impossibility of setting the position and the speed of an electron. The probabilistic nature of Quantum Mechanics' laws put great limits to the knowledge degree of an atomic system. After the reading of the paper "Quanten theorie des einatomige Idealen Gases II" by Einstein (February 9, 1925), and the analysis of De Broglie's work, the physicist Erwin Rudolf Josef Alexander Schrödinger, observed that the waves satisfied constraints similar to those imposed by the quantization conditions of Bohr. This foresaw the possibility of treating the electrons as "vibrating strings", and the oscillation modes as in relation to their energies.

In 1926, he presented a series of four articles entitled "Quantisierung Eigenwertals Problem I, II, III and IV" (3), subsequently collected in a December compendium ("An undulatory theory of the mechanics of atoms and molecules"). In his work, Schrödinger showed a quantum correlation among particles, and observed the entanglement phenomenon quantum (see Fig. 1) as an integrant part of the wave function [4]. However, he used the term "entanglement" for the first time in 1935, in the review article by Einstein, Podolsky and Rosen (EPR). Schrödinger definition was the following:

"When two systems, of which we know the states on the basis of their respective representation, undergo a temporary physical interaction due to known forces that act between them, and when, after a certain period of mutual interaction, the systems separate again, not we can describe them as before the interaction, i.e., endowing each of them with its own representation". (Proceedings of the Cambridge Philosophical Society, 1935, 31, p. 555).

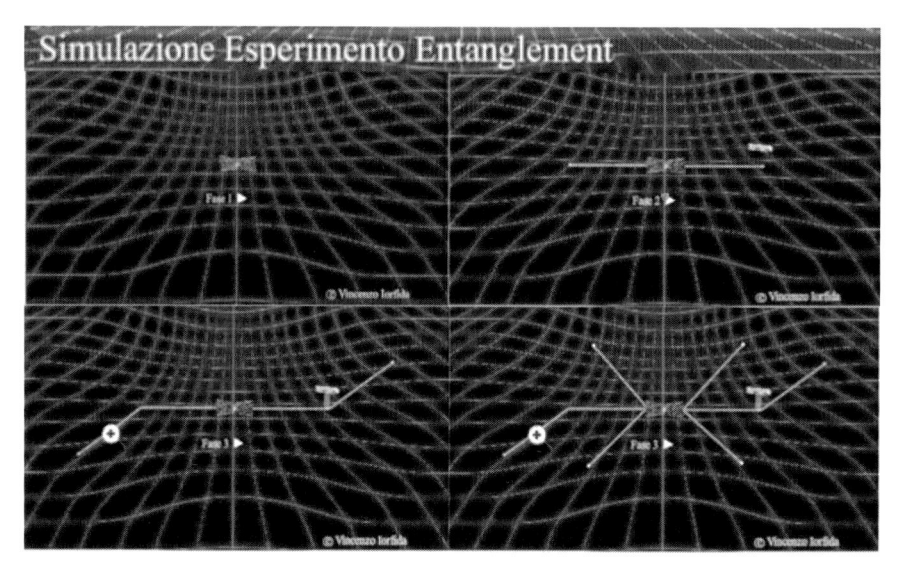

Figure 1. This picture shows four fundamental moments of a simulation on the quantum phenomenon of entanglement. The simulation was created in a multimedia environment with an interactive functionality, in order to provide the opportunity to learn the quantum phenomenon in an instructional environment.

According to Einstein quantum mechanics was an incomplete theory because it resulted in an instantaneous action at a distance and, on 15th May 1935, he published the article "Can be considered quantum-mechanical description of physical reality complete?" with his colleagues Boris Podolsky and Nathan Rosen. Through an ideal experiment, he highlighted how quantum theory violated the locality principle (4) and, in the conclusions of the work, affirmed: "*Although we have thus shown that the wave function does not provide a full description of physical reality, we left open the question of whether this description exist or not. We believe, however, that such a theory is possible*" (this experiment is known as the EPR paradox [5], [6]). In response to Schrödinger, he published "*Die gegenwärtige Situation in der Quantenmechanik*" on the 29th November, 1935. In this work he addressed the problem of the EPR paradox by declaring that the systems in a state of "entangled" imply a substrate of "non-separability" in the quantum description (holistic view). If completeness was claimed against the EPR, the non-separability had to be admitted: interacting systems were entangled, and the retrieval of individuality required the use of the collapse of the wave function. In 1964, John Stewart Bell published the article "On the Einstein - Podolsky - Rosen Paradox", by which he demonstrated how all the local theories were incompatible with quantum mechanics. He stated that "no physical theory of local hidden variables can reproduce the predictions of quantum mechanics". Moreover, he proved that in the local theories there is a restrictive upper limit, precisely defined by the Bell inequalities (their experimental assessment indicated the procedure to overcome the EPR paradox). Relevant experiments, including studies on the correlation of two photons (1982), led the French physicist Alan Aspect to affirm that "the analysis made by NHD Bohr is correct with respect to Einstein's interpretation". Such experiments continued to the present day with the award of the Nobel Prize for Physics in 2012 to Serge Haroche and David J. Wineland, "to have provided revolutionary experimental techniques that allow the measurement and manipulation of individual quantum systems" through the creation of entangled states, which represent a powerful tool for detection and manipulation in a controlled process.

THE ARTISTIC STRUCTURE

In describing the most interesting twenty-first century discovery in the field of quantum (i.e., the entanglement phenomenon), arts and science were parts of an empathic and dynamical synchronization process, revolutionizing the techniques and topics of the period. It is incontestable that the general meaning of art transformed over time. The evolution of quantum art in the twenty-first century, in all its many facets, played a key and innovative role, because it allowed the emerging of the relationships among the elements of a work. Unknowing the nature of the elements themselves, and detaching them from the (not absolute) concept of time and space, these relationships were not conformed to the principles of Euclidean geometry [7]. Through the structuring of new reference models and a close synergy with scientific imagination, the quantum art came to the interpretation of the vision of a new world (the subatomic one). In a world where physical processes are discontinuous and infinitesimal (represented by quantum jumps), the concept of information and of its simultaneous transfer, generated a non-linear behavior due to a complex system. In this view, the Quantum Art relied on studies that can convey emotions generating evolutionary interpretations of the microscopic world (which is not perceptible in another way), through a complex evolutionary process within the system in which the observer is an inseparable component.

The representation of this new artistic complex structure, transmitting and capturing innovative concepts, was integrated with the change of scientific and technological thought, through a state of continuous and evolutionary feedback, towards new figurative and dynamic models.

The constraint to the materialistic sensitivity provided limits to understand the world of the infinitely small, like that of sub-atomic particles. In fact, it is a world of which we have no direct experience because if it is difficult to observe through holographic patterns [8]: "You will never find the truth if you're not willing to accept even what you do not expect" (Heraclitus). The observer's role in quantum mechanics was not different from that of the artist: it was not static or stereotyped by elements belonging to the classical culture and describing everyday life. It was dynamic, chaotic and non-deterministic. Therefore, it was part of the system in dealing with a state of entanglement. Hence, it was explained the artist's search, who consciously does not reject the condition of "complexity", of new figurative theories representing the "delocalized reality" through the guiding principles of entanglement. This latter was difficult to represent through language and through the description of what occurs with non-deterministic character, like Quantum Theory. The use of computer technology through virtual procedures had a fundamental role in the quantum art for conserving the indeterminism of its imagery, regarding the relevant quantum reality, in a complex and interactive representation timeless coded. According to Schopenhauer, contemporary art was due not only to creative ideas, but it is supported by continuous research and a careful description representing a unperceivable and evolving part of the Universe. According to Schopenhauer, "representation is what we see, it has no objective foundation, then what we believe to be reality is a mere deception". However, Einstein believed that there was no space for subjectivity in science. Undoubtedly Salvador Dali was one of the major artists who shared the evolution of science in art [9], [10]. In fact, he was also interested in scientific exploration and took into account the concepts of Quantum Mechanics, presenting a work of 1954, "The Disintegration of the Persistence of Memory" (see Fig. 2). In this work, he represented the breaking down of memory into pieces, in honor to the new frontiers of science. In 1958 he wrote a "Manifesto Antimaterico: During the surrealist period I wanted to create the iconography of the interior world and the world of the wonderful designed by my father Freud. Today, the exterior world and that of physics has exceeded that of psychology. Today my father is Dr. Heisenberg".

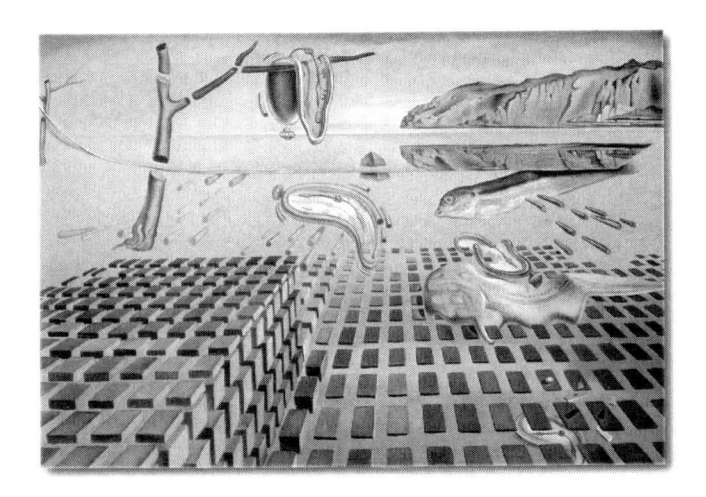

Figure 2. The Persistence of Memory (1931).

Figure 3. The Disintegration of the Persistence of Memory (1954).

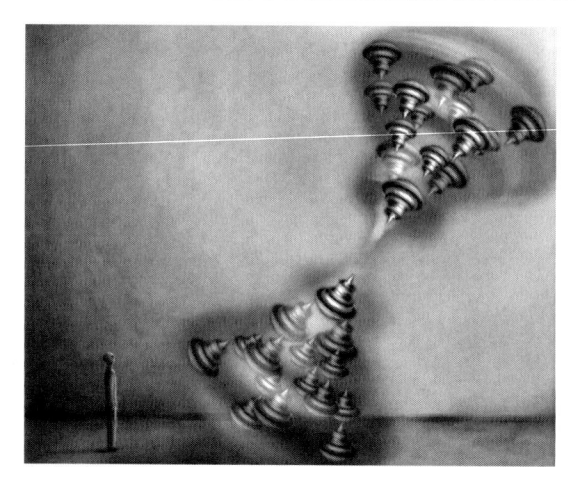

Figure 4. The quantum world, A. Pisani & V. Iorfida - Mathematical Art Galleries (2013).

CONCLUSION

The interpretation of the mathematical formalism of quantum mechanics foresees multiple interpretations. The most famous can be summarized in four areas: the Copenhagen interpretation, that of many worlds, of field guide, and the theory of spontaneous collapse - in addition to the more "revolutionary", linked to the concept of bayessian probability (5) with QBism acronym (6). This last theory does not support an objective reality but it is used mathematically by an observer to make the best decision on possible choices in the quantum world. The whole scientific communication and divulgation is fundamental, because a particularity could lead to a personal misperception, with a consequent distortion of the conveyance. Scenarios proposed by Quantum Art are not suggestive scenery, but they have an educational and scientific importance, that is not always easy to convey, because they are imperceptible in the macro world where we live and interact. In fact, the multidisciplinary instructional contribution of art is related to its scenarios (and not only from the information technology point of view). This is an important characteristic for a theory that it is not completely defined. According to Planck, intuition, direct perception or imagination are important factors for the progress of science. He stated that "the scientific researcher must have a very open intuitive imagination, because new ideas do not arise from deductions, but from an artistic creative imagination".

For this reason I believe that a interactive multimedia environment, and in particular the one I proposed, can explain a quantum phenomenon in a simple way. It can represent the entanglement without make it trivial, according to an innovative and educationally attractive approach crossing all disciplines (see Fig. 4).

REFERENCES

[1] Sardanashvily, G. (1995). Generalized Hamiltonian Formalism for Field Theory: *Constraint Systems*. World Scientific.

[2] D'Agostino, S. (2001). *A History of the Ideas of Theoretical Physics: Essays on the Nineteenth and Twentieth Century Physics*. Springer.

[3] Gamba, A., & Schiera, P. (2005). Fascismo e scienza. Le celebrazioni voltiane e il Congresso internazionale dei Fisici del 1927 [*Fascism and science. The Volta celebrations and the International Congress of Physicists of 1927*]. Bologna: Il Mulino.

[4] Audretsch, J. (2005). *Entangled Systems: New Directions in Quantum Physics*. Wiley-VCH.

[5] Afriat, A., & Selleri, F. (1999). *The Einstein, Podolsky and Rosen Paradox*. New York and London: Plenum Press.

[6] Bell, S. (1964). *On the Einstein Podolsky Rosen paradox. Physics 1.*

[7] Iorfida, V. (Inc. 2013). *Algebraic Curves and Varieties in XX Century Art. Chaos and Complexity Letters*. Nova Science Publishers.

[8] Talbot, M. (1992). L'universo olografico [*The holographic universe*]. New York: Mappae Perennial.

[9] Descharnes, R., & Descharnes, N. (1993). *Salvador Dalí*. New York: Konecky & Konecky.

[10] Klein, A. G. (2006). *Salvador Dalí*. United States: ABDO.

In: Chaos and Complexity in the Arts …   ISBN: 978-1-53612-995-3
Editors: N. Sala and G. Cappellato   © 2018 Nova Science Publishers, Inc.

Chapter 4

INVESTIGATING COMPLEXITY IN FOGGY PAINTINGS BY CNN-BASED TECHNIQUES

Arturo Buscarino[1], Luigi Fortuna[1], Mattia Frasca[1]*,
Angelo Lamia[1] and Maria Gabriella Xibilia[2]
[1]University of Catania, DIEEI, Catania, Italy
[2]University of Messina, DiSIA, Nuova Panoramica dello Stretto,
Messina, Italy

ABSTRACT

Aim of the paper is to deeply investigate the role and the characteristics of fog in various paintings by famous artists of different art movements, in which the presence of fog significantly affects the visual experience. To do this a new nonlinear signal processing technique, able to remove fog from color pictures exploiting optical properties of fog effect, is introduced and its implementation on a Cellular Nonlinear Network (CNN) is described.

Based on the assumption that, if fog is a real element of the natural scenario, then the artist can catch its optical effect in the artwork, the proposed methodology is used to investigate whether the fog is a natural element or it has been artificially added by the artist to express its own feelings.

A further analysis has been carried on to establish if the fog in some particular paintings may play the same role of noise in stochastic

---

* E-mail: mfrasca@diees.unict.it

resonance allowing to enhance some features which cannot be distinguished in absence of fog.

Keywords: cellular nonlinear networks, stochastic resonance, art, paintings, fog effect

1. INTRODUCTION

In some places the fog is still clean and the trees and houses, in the evening, seem objects without thickness, with the lightness of a vision.

(M. Soldati)

In this paper, the role of fog in artworks is investigated by using a new method based on Cellular Nonlinear Networks (CNNs). In particular, paintings in which the presence of fog is a characterizing element are considered. In this framework, fog can be seen as an added noise which may represent an enhancing element for the painting, allowing the onset of specific emotions in the observer [1, 2].

The possibility to obtain beneficial effects from noise injection is a now well established result, in several scientific fields [3]. However, the idea of investigating the role of noise in figurative arts has not been yet studied by means of image processing techniques. Under this perspective, this paper introduces a new nonlinear method which exploits the physical properties underlying the optical effects of fog, allowing to derive insights on the role of fog in paintings and to answer to some questions that are important for art criticism.

The proposed approach is based on a new algorithm implemented on CNN, parallel analog programmable devices widely used in image processing [4]. CNNs have been already used by the authors to investigate the artworks of specific art movements, namely kinetic art [5], or to understand the mind processes on which shape generation in modern art is based [6]. In [5], CNNs are used to reproduce dynamical emerging patterns which are recurrent subjects in modern Visual Art, like the Alberto Burri's Cretto or the Giacomo Balla artworks. The shapes characterizing paintings by Salvador Dalì, Robert Delaunay, and Juan Mirò are studied in [6] considering the emergent behavior of three dimensional CNNs.

In this paper, a different use of CNNs is proposed, since they are not used to reproduce artworks peculiarities but to implement the optical model of fog effects. This approach allows to discriminate if fog in paintings is in accordance to physical laws, and therefore if it was really present in landscapes painted from life or it was added by the artist for a specific aim.

The representation of fog has been considered under different perspectives in figurative art: from a rational approach, through which fog is represented by means of its physical features, towards an imaginative representation, used to enhance the emotional effect on the observer. Its ability to cover and uncover elements, altering shapes and dimensions, makes fog the vehicle of a new vision in which the artist can express his feelings and emotions.

The Italian art critic Achille Bonito Oliva [2] defines fog in contemporary art as "That state of necessary unawareness which the artist has to encompass to really assert his own identity [...]. Fog is metaphorically necessary, like drowsiness, allowing the wandering of the artist mind to the realization of new experience more than new experiments. The observer is forced to a state of stress and discomfort leading to a total concentration".

Fog is also associated to macabre images enhancing the sadness of particular landscapes and suggesting the temporary of life or the "mal de vivre". The observer will, therefore, imagine what is veiled by fog, ensuing his own sensitivity. Furthermore, fog seems to induce similar reactions in all humans [1], regardless their own life experiences, and painters have often exploited these effects to guide the observer towards peculiar emotions.

However, to infer the hidden aim of the artist when representing the fog in his paintings is not a simple task. In fact, two aspects have to be considered, the specific painter art movement and the geographical area in which the painter acts. Moreover, it should be inferred whether the artist represented a real foggy landscape or intentionally introduced fog to stimulate a particular emotive response of the observer.

The algorithm proposed in this paper allows to eliminate fog from paintings on the basis of its physical features. In details, a suitable model of optical effects of fog on human vision has been considered and used to implement an efficient image processing algorithm which is able to revert the fog effects removing them. The introduced algorithm has been implemented exploiting the parallel processing capabilities of Cellular Nonlinear Networks, and then used to evaluate the role of fog in visual art, and thus, to help in pursuing the distinction between real fog and imaginative fog in paintings.

The paper is organized as follows: in Section 2 the physical model of fog effect is discussed, in Section 3 the proposed methodology and its implementation on CNN are reported showing the algorithm performance on photos of real foggy landscapes. Our analysis on a set of selected paintings characterized by the presence of fog is reported in Section 4. Finally, the concluding remarks are drawn in Section 5.

2. OPTICAL MODEL OF FOG EFFECT

Fog is a natural phenomenon characterized by the condensation of water vapor near the Earth's surface, resulting in the decrease of visibility caused by suspended water droplets which modifies the optical properties of air.

The occurrence of fog needs that air is supersaturated of water vapor and this can happen in two different, often concurring, cases, i.e., when a mass of water is subjected to an intense evaporation, or when water vapor condensate in correspondence of a sudden temperature decrease.

The radiation coming from the sun interacts with suspended water droplets which cause light scattering. Therefore, an observer will be reached both by the direct radiation coming from the observed object and by that diffused by all the droplets laying between the object and the observer. Thus, the atmosphere acts like a source of light giving rise to the so-called *airlight* phenomenon. Airlight is caused also by a number of different natural light sources like the skylight, or the light reflected by the ground, or by the water. Furthermore, diffusion caused by droplets attenuates the beam of light coming from the observed object, since it is scattered in different directions.

The attenuation of light beams together with airlight causes chromatic effects which modify the perception of the observer. A model describing the total fog effect on colors, as perceived by digital color image acquisition tools, is the dichromatic atmospheric scattering model introduced in [7, 8]. The dichromatic model takes into account both attenuation sources, i.e., airlight and scattering, and can be written as:

$$
\begin{aligned}
\mathbf{E} &= p\hat{\mathbf{D}} + q\hat{\mathbf{A}} \\
p &= Re^{-\beta d} \\
q &= E_{\infty}\left(1 - e^{-\beta d}\right)
\end{aligned}
\tag{1}
$$

where $\mathbf{E} \in \Re^{3}$ represents in the RGB space the observed color of a scene point in presence of fog, $\hat{\mathbf{D}} \in \Re^{3}$, $\hat{\mathbf{A}} \in \Re^{3}$, are unit vectors representing the color of the same scene point as seen in clear weather conditions, and the airlight color, respectively, p is the magnitude of direct transmission and q is the magnitude of airlight; E_∞ is the sky brightness, R is the radiance of the scene point on a clear day, β is the scattering coefficient of the atmosphere, and d is the distance of the scene point from the observer.

The dichromatic model can be applied to remove the effects of fog by suitably choosing its parameters [7]. In this paper, the dichromatic model is implemented in a CNN which processes grayscale images. For this reason the RGB image is split in the three channels obtaining three grayscale images which can be processed separately by using the contrast or monochromatic model derived from the dichromatic model in Eqs. (1), to mathematically identify the intensity E of a scene point as recorded by a monochromatic camera [7]:

$$E = Re^{-\beta d} + E_\infty\left(1 - e^{-\beta d}\right) \tag{2}$$

This model can be used to recover an abstract image without fog if R is found by inverting relation (2). The CNN algorithm, which will be discussed in details in the next section, implements the inverse relation of the monochromatic model.

Furthermore, the model is based on single-scattering, i.e., turbulent blurring or scattering introduced by pollutants are not considered. This means that only the scattering introduced by particles forming fog will be removed through our algorithm.

3. REMOVING FOG BY CNN

The algorithm for the elimination of fog effects has been realized by using the classical CNN-Universal Machine (CMM-UM) paradigm [9], a programmable analog array computer based on the concept of Cellular Nonlinear Networks.

The first assumption in the CNN approach is that each CNN cell elaborates one pixel of the image to be processed and evolves according to the dynamical model described in [10] whose equations are:

$$\dot{x}_{ij}(t) = -x_{ij}(t) + \sum_{c(k,l)\in N_r(i,j)} A(i,j;k,l)y_{kl}(t) + \sum_{c(k,l)\in N_r(i,j)} B(i,j;k,l)u_{kl}(t) + I$$
$$y_{ij} = 0.5 * \left(\left|x_{ij} + 1\right| - \left|x_{ij} - 1\right|\right) \tag{3}$$

where A and B are called feedback and control templates, I is a bias term,

$N_r(i,j)$ represents the set of r-neighbor cells of cell (i,j), and u_{ij} and y_{ij} are the input and output of the cell (i,j), respectively. Choosing a set of templates (A, B, I), the CNN can be programmed to perform a specific image processing task.

In order to perform complex tasks, different sets of templates can be cascaded, implementing the so-called CNN-UM. In this case, the CNN device is programmed through an algorithm in which the output of each step is the input of the following step.

In this section, the CNN-UM algorithm reproducing the monochromatic model for fog removal will be described.

Since our aim is to remove fog from images according to optical laws, the first step is to invert the monochromatic model deriving the radiance R from Eq. (2):

$$R = \left[E - E_\infty\left(1 - e^{-\beta d}\right)\right]e^{\beta d} \qquad (4)$$

We assume for E_∞ the RGB component of the white color, i.e., $E_\infty = 255$. The atmospheric diffusion coefficient β assumes different values in correspondence of the fog density: higher values of β allow to remove dense fog and viceversa. Hence, β can be a space-variant quantity.

RGB color images are processed splitting the three channels and applying the same CNN algorithm to each gray-scale image. Each cell therefore represents a pixel of the image and it is initialized with the corresponding normalized gray scale value (i.e., in the CNN notation, -1 represents white, while 1 represents black).

The CNN implementing Eq. (4) is programmed through an algorithm consisting of two steps, each associated to a specific set of templates. In the following, the role of each step will be outlined and explained through the elaboration of a sample painting, "La nebbia sale" (The fog is rising) by the contemporary Italian painter Alfio Cioffi shown in Figure 1(a). The image normalized according to the CNN notation is reported in Figure 1(b) and then split in the three RGB channels.

The input of the CNN is a gray scale image built according to:

$$U_{i,j} = \frac{E_\infty}{2}\left(1 - e^{-\beta d_{i,j}}\right) \qquad (5)$$

where $d_{i,j}$ represents the distance of pixel *(i,j)* from the imaginary observer plane. For the painting considered in this example, the image shown in Figure 2(a) has been used as input.

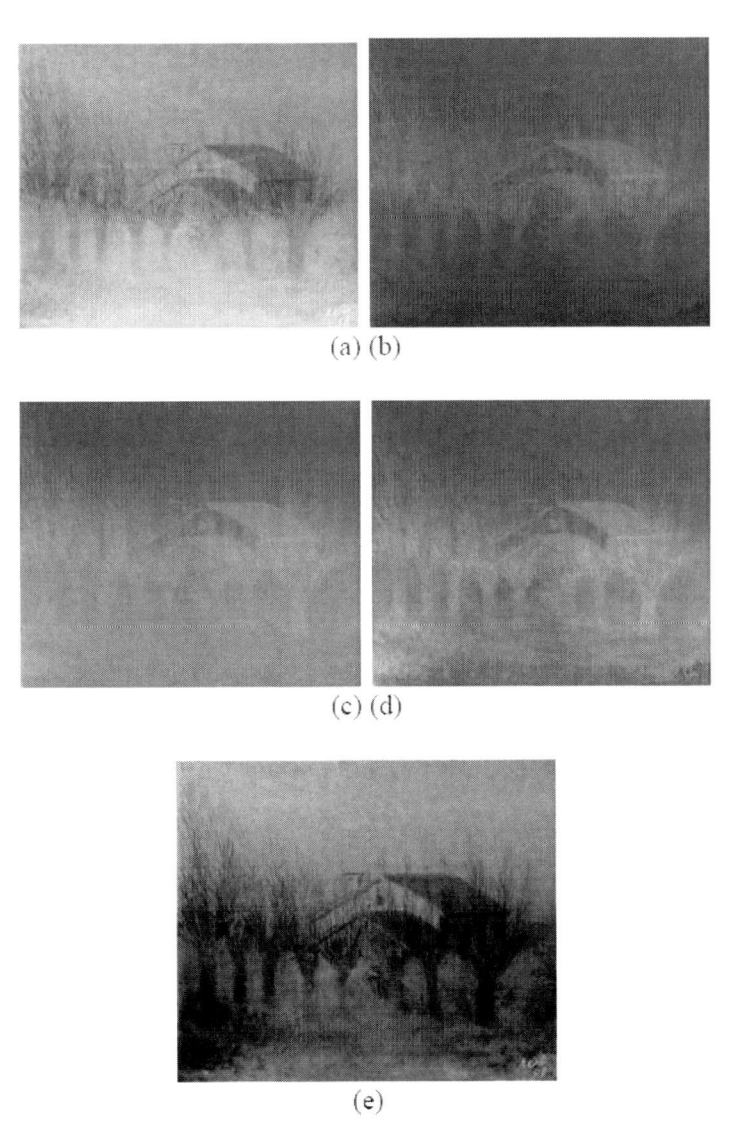

(a) (b)

(c) (d)

(e)

Figure 1. (a) Alfio Cioffi - La nebbia sale (The fog is rising), 2006, oil on canvas board, 40x50, (b) normalized image used as initial conditions for the first step, (c) output of the first step, (d) output of the algorithm, (e) re-normalized image in which fog has been removed.

Figure 2. Images used by the CNN algorithm as (a) input of the first step and (b) initial condition for the second step.

The objective of the first step is to perform the algebraic sum $E - E_\infty\left(1 - e^{-\beta d}\right)$. The following set of space-variant templates allows, in fact, to obtain the sum $X_{i,j}(0) - |U_{i,j}|$.

$$A = \begin{pmatrix} 0 & 0 & 0 \\ 0 & -1 & 0 \\ 0 & 0 & 0 \end{pmatrix}; \ B = \begin{pmatrix} 0 & 0 & 0 \\ 0 & 2 & 0 \\ 0 & 0 & 0 \end{pmatrix}; \ I = 2X_{i,j}(0) \tag{6}$$

The image obtained merging the three outputs in the RGB space of the first step after 20 iterations is reported in Figure 1(c) and it is used as input image of the second step.

In this step, the initial conditions are set as a gray scale image built according to:

$$X_{i,j}(0) = \frac{E_\infty}{2}\beta d_{i,j} \tag{7}$$

In this case, the image shown in Figure 2(b) has been used as initial conditions.

The following set of space-variant templates has been applied to perform the product $U_{i,j} e^{X_{i,j}(0)}$:

$$A = \begin{pmatrix} 0 & 0 & 0 \\ 0 & -1 & 0 \\ 0 & 0 & 0 \end{pmatrix}; \ B = \begin{pmatrix} 0 & 0 & 0 \\ 0 & 2e^{X_{i,j}(0)} & 0 \\ 0 & 0 & 0 \end{pmatrix}; \ I = 0 \tag{8}$$

The image resulting after 200 iterations is reported in Figure 1(d) which has to be re-normalized to obtain the final image reported in Figure 1(e). As it can be observed, the fog effect has been removed from the painting revealing more details of the landscape. The complete CNN algorithm is schematically summarized in Figure 3.

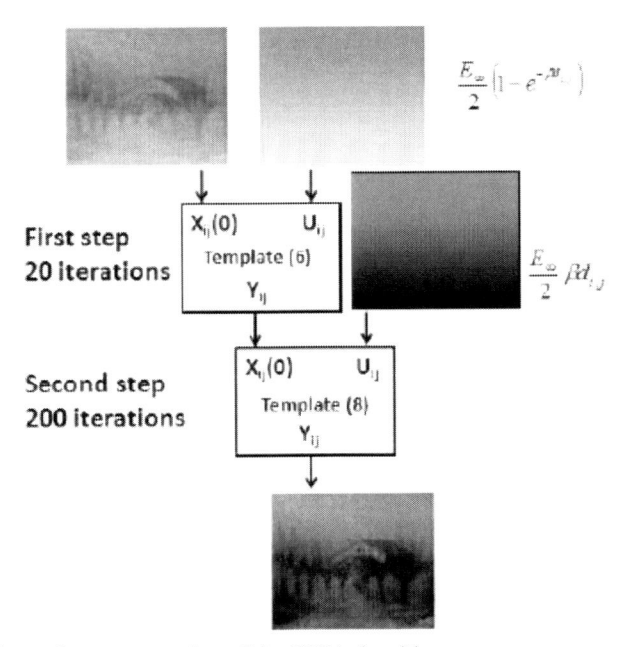

Figure 3. Schematic representation of the CNN algorithm.

The efficiency of the proposed algorithm in removing fog effect has been tested on two color pictures. The results can be observed in Figure 4 and in Figure 5. In more details, in Figure 4 the optical effect of fog has been removed unveiling the hidden details of the picture. In Figure 5, a landscape with chimneys of a refinery plant is shown. It can be observed that the algorithm is effective in removing only fog effect, while, as expected, the smog produced by the refinery is still visible. In the next Section, the algorithm will be applied on several paintings in order to investigate the fog effects.

4. RESULTS

As a first example, we consider another painting by Alfio Cioffi reported in Figure 6. In particular, Figure 6(a) shows the original painting, while Figure 6(b) the result of the CNN processing.

It can be observed that fog effect has been removed indicating that the represented fog respected the physical laws underlying the optical effect of real fog. Even if fog has been removed, no further details are unveiled after fog removal. This can be an indication that fog was really present during the artwork production, i.e., the subject has been painted from life.

Figure 4. Landscape with fog.

Figure 5. Landscape with fog and pollutants.

The same conclusion can be drawn for the famous painting "Felsenlandschaft im Elbsandsteingebirge" (Rocky landscape in the Elbe Sandstone Mountains), reported in Figure 7, by Caspar David Friedrich (1774-1840). Friedrich was instrumental in transforming landscape in art from a backdrop subordinated to human drama to a self-contained emotive subject. In this painting the fog removal has been effective; this can lead to the conclusion that the fog was a real element of the landscape.

A different example is represented by the painting "La battaglia di Pavia" (Battle of Pavia), shown in Figure 8, by Gherardo Poli (1676-1739), artist of landscapes in which real elements are transformed in fictitious representations. In this case, the subject is an event occurred in 1525.

After the CNN processing, several hidden particulars emerge from fog, pointing out that fog has been added artificially, even if its effects are carefully reproduced. Furthermore, it can be observed that smoke over Pavia is preserved after the fog removal.

As a further example, the famous painting "Der Wanderer über dem Nebelmeer" (Wanderer above the Sea of Fog), shown in Figure 9, is considered. In this case, the CNN processing does not allow to remove fog, indicating that it does not respect the optical laws of real fog. Fog is an additional fictitious element intentionally introduced by the artist which, according to [11], wanted to use the wanderer's gazing into the sea of fog to express a sort of self-reflecting feeling.

Figure 6. Alfio Cioffi - L'argine (The riverbank), 2006, oil on canvas board, 35x50.

In the examples discussed above, fog has been considered as a representation of either a real element or painter feelings. However, when fog is intentionally added, one may wonder if the artist wanted to enhance a specific peculiarity or some particulars of the painting. In this sense the fog acts as does the noise in some nonlinear systems exhibiting stochastic resonance. In fact, noise is generally considered as an unwanted effect, but in the last years experimental results have suggested that noise addition might improve the performance of nonlinear systems. Stochastic resonance is a quite ubiquitous phenomenon that arises in a number of cases, spanning from engineering to biology, and shows that a beneficial effect can result from the interaction between noise and signals. Hence, the intentional addition of noise in a system, as fog in a painting, allows the emergence of hidden properties and features [3, 12].

Figure 7. Caspar David Friedrich - Felsenlandschaft im Elbsandsteingebirge (Rocky landscape in the Elbe Sandstone Mountains), oil on canvas, 91x74, Wien, Österreichische Galerie.

Figure 8. Gherardo Poli - La battaglia di Pavia (Battle of Pavia), oil on canvas, 84x128, Musei Civici di Pavia.

It is quite interesting regarding this latter case that evidence has been given on the existence of stochastic resonance in biological systems, especially as regards the visual perception [13].

Figure 9. Caspar David Friedrich - "Der Wanderer über dem Nebelmeer" (Wanderer above the Sea of Fog), oil on canvas, 98x75, Kunsthalle Hamburg.

Figure 10. Claude Oscar Monet - from the "London House of Parliament Series".

This phenomenon has been revealed in our analysis of one of the Claude Oscar Monet (1840-1926) paintings of the famous "London House of Parliament Series". Comparing the original painting to the processed image, both shown in Figure 10, it can be observed that the fog effect has been removed. However, fog elimination has also make less relevant some particulars of the paintings like the small boat, circled in Figure 10, which in the original painting stands out in the fog.

CONCLUSION

In the paper, a new tool designed to remove fog effects from pictures according to the optical properties defined in the so-called dichromatic model has been proposed. The proposed approach is based on the implementation of the inverse dichromatic model on a CNN-UM, an analog parallel processor which can be suitably programmed in order to perform specific image processing tasks. This tool can be very useful in gaining new insights on the role of fog in paintings, and therefore in investigating painters intents. The use of fog in paintings, in fact, may assume a relevant role since allows the painter to convoy the observer towards a specific mood. Albeit this study cannot give definitive conclusions, since this cannot leave aside a deeper knowledge of the painters, their art movements and the scene in which they operate, the proposed algorithm can be considered as a useful tool to discriminate whether or not fog has been painted according to physical laws. Interesting results have been obtained by processing a set of paintings by different artists and art movements, in some cases unveiling hidden features and particulars. Furthermore, some of the results shown open the way to possible interpretations of the fog effect in terms of a stochastic resonance phenomenon, in which the fog plays the role traditionally played by the noise in the enhancement of a feature of the system.

REFERENCES

[1]   S. Zeki, *Inner Vision: an exploration of art and the brain*, Oxford University Press, USA, 2000.

[2]   Bonito Oliva, Il tallone di Achille. *Sull'arte contemporanea (Campi del sapere)* [Achilles' heel. On contemporary art (Fields of knowledge)], Feltrinelli, 1988.

[3]   Andò, Stochastic Resonance: *Theory and Applications*, Springer, Boston, 2000.

[4]   L.O. Chua and T. Roska, *Cellular neural networks and visual computing - Foundations and applications*, Cambridge University Press, 2005.

[5]   M. Bucolo, A. Buscarino, L. Fortuna, M. Frasca and M.G. Xibilia, From Dynamical Emerging Patterns to Patterns in Visual Art, *Int. J Bifurcation and Chaos* 18 (2006), pp. 51–81.

[6] P. Arena, M. Bucolo, S. Fazzino, L. Fortuna, and M. Frasca, The CNN Paradigm: Shapes and Complexity, *Int. J. Bifurcation and Chaos* 15 (2005), pp. 2063–2090.

[7] S.G. Narasimhan, and S.K. Nayar, Vision and the Atmosphere, *Int. J. of Comp. Vision* 48 (2002), pp. 233–254.

[8] S.G. Narasimhan, and S.K. Nayar, Interactive (De)Weathering of an Image using Physical Models, in *Proc. of IEEE Workshop on Color and Photometric Methods in Computer Vision*, October 2003.

[9] T. Roska, and L.O. Chua, The CNN universal machine: An Analogic Array Computer, *IEEE Trans. Circuits and Systems* II 40 (1993), pp. 163–173.

[10] L.O. Chua, and L. Yang, Cellular neural networks: applications, *IEEE Trans. Circiuts and Systems* 35 (1988), pp. 1273–1290.

[11] M.E. Gorra, *The bells in their silence: Travels through Germany*, Princeton University Press 2004.

[12] G. Napoli, and M.G. Xibilia Soft Sensor design for a Topping process in the case of small datasets, *Comp. and Chem. Eng.* 35 (2011), pp. 2447-2456.

[13] E. Simonotto, M. Riani, C. Seife, M. Roberts, J. Twitty, and F.Moss, Visual Perception of Stochastic Resonance, *Phys. Rev. Lett.* 78 (1997), pp. 1186–1189.

In: Chaos and Complexity in the Arts …   ISBN: 978-1-53612-995-3
Editors: N. Sala and G. Cappellato   © 2018 Nova Science Publishers, Inc.

Chapter 5

FRACTALS IN ARCHITECTURE: HYPERARCHITECTURE AND BEYOND

Nicoletta Sala[*]

Accademia di Architettura di Mendrisio
Università della Svizzera Italiana, Svizzera (Switzerland)

ABSTRACT

During the centuries the architecture has followed the Euclidean geometry and Euclidean shapes. Thus the buildings had Euclidean aspects, but some architectural styles, for example the Baroque and the Hindu architecture, are informed by Nature, and much of Nature is manifestly fractal and complex.

The aim of this paper is to describe where the fractality appears in architecture and in urban organization, opening new opportunities in the virtual architecture and in the hyperarchitecture, too.

Keywords: architecture, euclidean geometry, fractals, liquid architecture, hyperarchitecture, hypersurface, transarchitecture, virtual architecture, virtual reality

---

[*] E-mail address: nicoletta.sala@usi.ch

1. INTRODUCTION

For many centuries architects found inspiration by the Euclidean geometry and by the Euclidean shapes (for example, triangles, squares, Platonic solids, and polyhedrons), and we are not surprise to observe that the buildings have Euclidean aspects, as shown in ancient Egyptian *El Giza Pyramids* (in Figure 1a) (Sala and Cappellato, 2003a). The presence of the Euclidian geometry and of the polyhedric shapes ensured the structural stability to the buildings. These shapes are also present in the architecture of the twentieth century. Swiss architect Le Corbusier, pseudonym of Charles-Edouard Jeanneret (1887-1965), one of the most important leading figure of modern architecture and urbanism, in the *Dominican Convent of Sainte Marie de la Tourette* (1957-1960, Eveux sur-Arbresie, France) used pure geometrical forms, like the polyhedrons. In particular, Le Corbusier adopted very simple and lean forms in his architectural solution, to reflect on the origins and the rules of the Dominican order (Sala and Cappellato, 2003a). The Figure 1b shows the vision of convent area that evidences its geometric characteristics.

(a)   (b)

Figure 1. (a) Egyptian El Giza Pyramids (b) Dominican Convent of Sainte Marie de la Tourette (Eveux sur-Arbresie, France.

On the other hand, some architectural styles are informed by Nature. Figure 2a shows a Vitruvius' draw that described the creation of the first Corinthian capital influenced by the shapes of the acanthus leaves. Figure 2b illustrates a Corinthian capital.

Francesco Borromini (1599-1667) found inspiration in the Nature, in particular observing the shells. He used the octagons, the Greek crosses and other shapes for the coffering of the dome of *San Carlo alle Quattro Fontane*

(1638-1641). The Figure 3a illustrates the valve lattice of the shell (*Cakadia*) which provided the brunched co-ordinates that map out the Greek crosses and the octagons, shown in Figure 3b, that Borromini used to cover the dome of *San Carlo alle Fontane*. The Figure 4 illustrates the dome interior where the ends of each lozenge and of each rhombus are unequal, the upper half of each octagon is smaller than the lower half, and the top of the upright in each Greek cross is shorter than the bottom of the lower part of the cross' upright (Hersey, 1999). In Figure 4, it is possible to observe the presence of two directional compressions, horizontal and vertical at the same time, over a (much shallower) dished plan. These reductions in smaller and smaller scales in the dome can be explained also using a fractal point of view (Sala and Cappellato, 2003b).

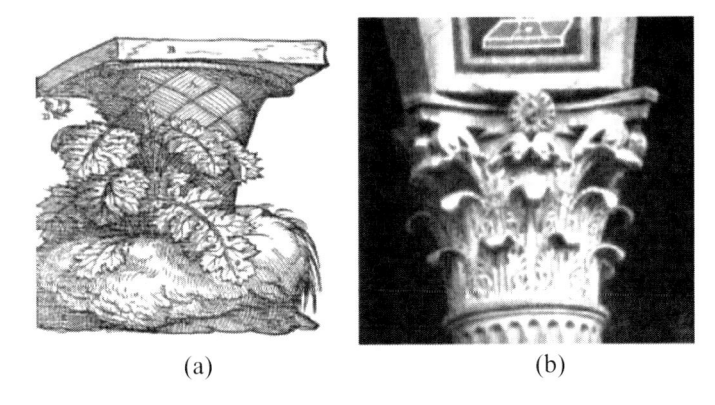

(a) (b)

Figure 2. (a) Vitruvius' draw that described the creation of the first Corinthian capital (b) Corinthiancapital.

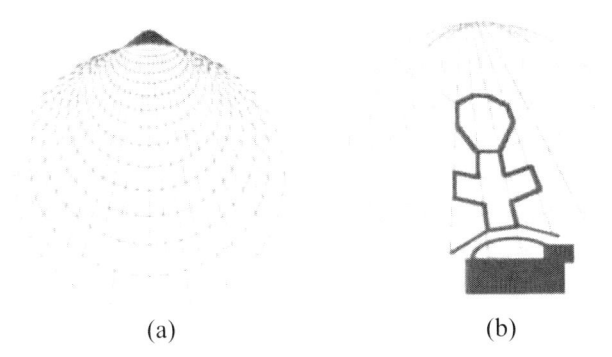

(a) (b)

Figure 3. (a) The valve lattice of *Cakadia*, the lattice used to map a detail in the (b) Borromini's dome (Hersey, 1999, p. 52).

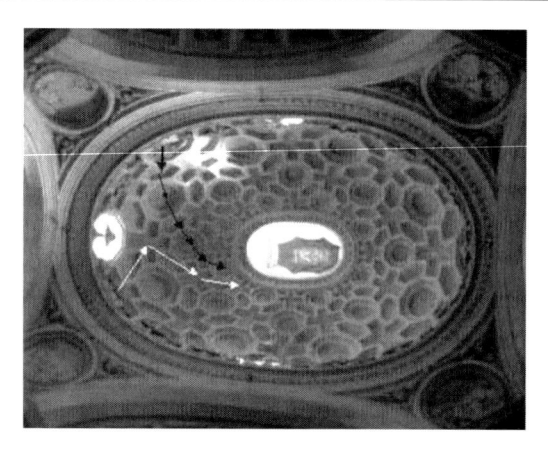

Figure 4. Dome of *San Carlo alle Quattro Fontane* (1638-1641, Rome).

Mario Botta affirmed: «*The nature must be part of the architecture like architecture must be nature part; the two terms are mutually complementary. The architecture describes the man's project, the life space organization and therefore is a work thought reason act. Thus, this is always a "dialogue" and a comparison with the nature. (La natura deve essere parte dell'architettura così come l'architettura deve essere parte della natura; i due termini sono reciprocamente complementari. L'architettura descrive il progetto dell'uomo, l'organizzazione dello spazio di vita e quindi è un atto di ragione, di pensiero, di lavoro. Proprio per questo è sempre "dialogo" e confronto con la natura)*». (Sala and Cappellato, 2003a, p. 12)

Nature has influenced the architects in different cultures and in different period, and modern architecture also mimics natural shapes (for instance, in the textures of Frank O. Gehry or in tree structures of Otto Frei).

(a)   (b)

Figure 5. (a) *Hemispheric* (Valencia) by Calatrava (Design: 1991-1995. Construction: 1996-1998). (b) An armadillo.

Spanish architect and engineer Santiago Calatrava has realized complex buildings and bridges. The shapes of his projects suggests a complex interpretation influenced by the observation of nature. Figure 5a shows the "Hemispheric", an amazing glass structure, realized in the "Ciudad de las Artes y las Ciencias" (Valencia, Spain), which evidences an interesting analogy with the shape of complex cuirass of the armadillo, as shown in Figure 5b (Sala, 2003b).

2. FRACTALS IN ARCHITECTURE

After the Mandelbrot's definition of fractal (1975), different interdisciplinar applications of this new geometry was realized. For example, in biology, in medicine, in computer science, in economy, in telecommunication, in religions and so forth. (Lindenmayer, 1968; Mandelbrot, 1975; Prusinkiewicz, 1986; Prusinkiewicz and Lindenmayer, 1990; Leland et al, 1993; Nonnenmacher, et al., 1994; Jackson, 2004; He et al., 2006; Sala, 2008). Obviously, some of these applications were in architectural field, but rigorous discussion of the link between fractals and architecture can lead to ambiguous territory, where the rules have been not already defined (Jencks, 1997; Salingaros, 1999, 2000, 2001; Ostwald, 2001; Sala and Cappellato, 2004; Capo, 2004, Sala, 2006).

Mandelbrot, Prusinkiewicz and Lindenmayer demonstrated that nature is manifestly complex and fractal (Lindenmayer, 1968; Mandelbrot, 1975, 1982; Prusinkiewicz and Lindenmayer, 1990). Therefore, we should not be so surprised to find complex and fractal components in architecture and the design, in particular for reproducing the patterns and the shapes present in Nature.

We reduced the difficulties dividing our fractal analysis in architecture in three parts. In the first we did a small scale analysis (e.g., to determine the single building shape, its degree of complexity and the generative process). The second part was oriented on a large scale analysis (e.g., to study the urban morphology, the urban growth and the urban development) (Batty, 1991, Batty and Longley, 1994, 1997; Frankhauser, 1994, 1997; Donato and Lucchi Basili, 1996; Sala, 2000; Saleri, 2008). In the third part we considered fractal algorithms to generate virtual architectures.

In agreement with Falconer's point of view (1990) we consider fractal sets F as follow (Falconer, 1990):

1. F has a fine structure, that is, details at an arbitrarily small scale;
2. F is too irregular to be described in a traditional geometric language, both locally and globally;
3. F often has some form of self-similarity, perhaps approximate or statistical;
4. Usually the "fractal dimension" of F (defined in one way or another) is greater than its topological dimension;
5. In the most part of the cases, F can be defined in a very simple way, perhaps recursively.

We also have classified the use of fractal geometry in architecture in two different ways: unconscious, when the fractal quality has been unintentional chosen for an aesthetic sense (for example, in the Hindu, in the Gothic and in the Baroque architectures) or conscious, when the fractal quality is, in every case, the result of a specific and conscious act of design (for example, in the contemporary architecture).

2.1. Fractals in the Small Scale Analysis

In the small scale analysis we have observed:

-   The box-counting dimension of a design, to determine its degree of complexity (Bovill, 1995; Sala and Cappellato, 2004, Burkle et al., 2004).
-   The building's self-similarity (e.g., a building's component which repeats itself in different scales) (Sala 2000; Sala, 2002; Sala, 2003, Sala and Cappellato, 2003b; Sala and Cappellato, 2004).
-   The Iterated Function System to understand the generative processes of the complex buildings (Sala and Cappellato, 2004; Sala, 2005).

2.1.1. Box-Counting Dimension

The box-counting dimension is connected to the problem of determining the fractal dimension of a complex two-dimensional image for its synthetic description. It is defined as the exponent Db in the relationship:

$$N(d) \approx \frac{1}{d^{D_b}}$$

(1)

where N(d) is the number of boxes of linear size d, necessary to cover a data set of points distributed in a two-dimensional plane. The basis of this method is that, for objects that are Euclidean, equation (1) defines their dimension. One needs a number of boxes proportional to $1/d$ to cover a set of points lying on a smooth line, proportional to $1/d_2$ to cover a set of points evenly distributed on a plane, and so on. Applying the logarithms to the equation (1) we obtain: $N(d) \approx -D_b \cdot \log(d)$.

The box-counting dimension can be produced using this simple iterative procedure:

- superimpose a grid of square boxes over the image (the grid size is s_1);
- count the number of boxes that contain some of the image ($N(s_1)$);
- repeat this procedure, changing (s_1), to smaller grid size (s_2);
- count the resulting number of boxes that contain the image ($N(s_2)$);
- repeat this procedures changing s to smaller and smaller grid sizes.

The box-counting dimension is defined by:

$$D_b = \frac{[\log(N(s_2)) - \log(N(s_1))]}{\left[\log\left(N\left(\frac{1}{s_2}\right)\right) - \log\left(N\left(\frac{1}{s_1}\right)\right)\right]} \tag{2}$$

where $1/s$ is the number of boxes across the bottom of the grid. We can apply the box-counting dimension in the architecture, too (Bovill, 1995). It is calculated by counting the number of boxes that contain lines from the drawing inside them. Next Figure 6 shows the procedure to determine the "complexity" of Wright's *Robie House* (1909) using the box count, and Table 1 resumes the data (Bovill, 1995).

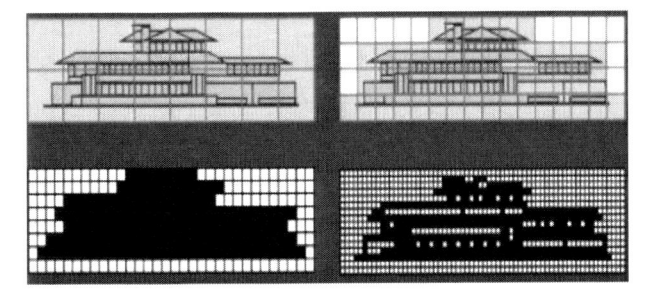

Figure 6. Box-counting method applied to the elevation of the *Robie House* (1909). Number of boxes counted, the number of boxes across the bottom of the grid, and the grid size.

Table 1. Data collected in the box counting process for *Robie House*

Box count	Grid Size	Grid Dimension (feet)
16	8	24
50	16	12
140	32	6
380	64	3

Three dimensions can be calculated. The first is for the increase in number of boxes with lines in them from the grid with 8 boxes across the bottom (24 feet) to the grid with 16 boxes across the bottom (12 feet). The next scanning range compares boxes that are 12 feet across with boxes that are 6 feet across, and the third scanning range compares boxes that are 6 feet across with boxes that are 3 feet across.

$$D_{(box,24'-12')} = \frac{[\log(50) - \log(16)]}{\left[\log(16) - \log(8)\right]} = \frac{(3.912 - 2.773)}{(2.773 - 2.080)} = \frac{1.139}{0.693} = 1.644$$

$$D_{(box,12'-6')} = \frac{[\log(140) - \log(50)]}{\left[\log(32) - \log(16)\right]} = \frac{(4.942 - 3.912)}{(3.466 - 2.773)} = \frac{1.03}{0.693} = 1.486$$

$$D_{(box,6'-3')} = \frac{[\log(380) - \log(140)]}{\left[\log(64) - \log(32)\right]} = \frac{(5.940 - 4.942)}{(4.159 - 3.466)} = \frac{0.998}{0.693} = 1.440$$

The Box-counting dimension of *Robie House* is: 1.440< Db <1.486.

Burkle et al. (2004) applied the same approach to determine the degree of the complexity in the Mesoamerican architecture, in particular in the pyramids (Burkle, Sala and Cepeda, 2004; Burkle and Sala, 2004). A pyramid is a series composed by different number of platforms of different measures. In detail, they analyzed in the Mesoamerican pyramids all the structure together, but by the other side, seeing that these buildings like boundaries and try to study their separated sequential segments in order to achieve to understand better the distinct aspects of the correlation functions describing the fractality. These geometric studies of monumental art and architecture had been done looking the pyramids like if they were Platonic bodies, cubes or tetrahedrons, and like Euclidean structures as rectangles, squares, triangles and circles. But the pyramids are irregular and complex buildings of massive and heavy volume.

The aim of this research was to study these structures trying to find out the patterns and designs and the forms into this complex geometry that appear to enclose a specific guide of information encode in them, and to decipher the possible interconnected nature of different reckoning systems. Data analyses on the fractality and the complexity in Mesoamerican architecture suggested that the architects of these pyramids tried to imagine some models observing the nature.

Basic symbolism representing the cosmos vision are present in the Mesoamerican pyramids related with earth, water and fertility, mountains and caves. All these symbols are the manifestations of a cult system that included not only the cosmology, but the complex mathematics involved in it and in the mythic and the ritual concepts (Burkle and Sala, 2004).

2.1.2. Self Similarity in Architecture

In mathematics, a self-similar object is exactly or approximately similar to a part of itself (i.e., the whole has the same shape as one or more of the parts). The self-similarity is a property by which an object contains smaller copies of itself at arbitrary scales. "Similar" means that the relative proportions of the shapes' sides and internal angles remain the same. Fractals can also be classified according to their self-similarity. As described by Mandelbrot (1982), this property is ubiquitous in the natural world (Mandelbrot, 1982). Oppenheimer (1986) used the term "fractal" exchanging it with self-similarity, and affirmed: «*The geometric notion of self-similarity became a paradigm for structure in the natural world. Nowhere is this principle more evident than in the world of botany.*» (Oppenheimer, 1986)

There are three kind of self-similarity:

- Exact self-similarity. The fractal is identical at different scales. This is the strongest kind of self-similarity.
- Quasi-self-similarity. The fractal is approximately (but not in exact way) identical at different scales. This is a less precise form of self-similarity. Quasi-self-similar fractals contain small copies of the entire fractal in degenerate and distorted shapes. This is the kind of fractals defined by recurrence relations.
- Statistical self-similarity. The fractal has statistical or numerical measures which are preserved across scales; instead of specifying

exact scales, at each iteration the scale of each piece is selected randomly from a set range. This is the weakest kind of self-similarity. Most common definitions of "fractal" imply this kind of self-similarity. Random fractals are examples of fractals which are statistically self-similar.

During our research we have also classified the use of self-similarity in architecture in two different ways: unconscious, when the fractal quality has been unintentional chosen for an aesthetic sense, and conscious, when the fractal quality is in every case the result of a specific and conscious act of design. The use of the conscious self-similarity in the modern architecture has been facilitated by the introduction of computers and their software tools in the process of design (Eaton, 1998; Sala, 2000; Sala, 2002). In the next sections we describe the self similarity in Hindu architecture, Oriental architecture and modern architecture.

Hindu Architecture

For over two centuries much of Asia has been dominated by Indian Hinduism as a religious, political and social force. Hindu Asia was constituted by the subcontinent of India, the peripheral sub-Himalayan valleys, the major part of mainland South-East Asia and the Indonesian archipelago. Most important artistic expression of Hinduism is the temple, which was thought to create a link between the world of man and the world of the gods. To understand the architectural shapes of the Hindu temples it is necessary to study the origins and the development of the civilization that conceived them. In older cultures the mountains represented the sacred sanctuaries around the world. In the Hindu experience the idea of the archetypal mountain of existence is mythologized in the cosmic mountain named Mount Meru (or Sumeru), the mythological center of all the physical, metaphysical and spiritual universes (Mitchell, 2000). To design the temples, Indian and Southeast Asian architects found inspiration observing the mountains. George Mitchell (1988) wrote:

> «*In the superstructure of the Hindu temple, perhaps its most characteristic feature, the identification of the temple with the mountain is specific, and the superstructure itself is known as a 'mountain peak' or 'crest' (shikhara). The curved contours of some temple superstructures and their tiered arrangements owe much to a desire to suggest the visual effect of a mountain peak*» (Mitchell, 1988, p. 69).

The fractal structure of the mountains has been researched and discussed by analysts, for instance to *develop a mechanism for generating a kind of fractal mountains based on recursive subdivision* algorithm for a triangle (Fournier et al., *1982)*. Thus, we are not surprised to denote that Indian and Southeast Asian temples and monuments exhibit a fractal structure. In fact, the towers are surrounded by smaller towers, surrounded by other smaller towers, and so on, for seven or more levels.

In these cases the proliferation of towers represents various aspects of the Hindu pantheon, and shows that it typically involves a multiple set of ideas (Jackson, 1999*). Hindu traditional architecture has more symbolic meanings than the architecture of other cultures, in particular it is highly articulated. Hindu temple is oriented to face East, the direction where the sun rises on the darkness, and its design includes an image of a Cosmic Person in yogi-like position. It also includes the archetype of the cosmic mountain, the cave of sacred inner mystery, and other imageries (the connection between earth and heaven, the fertility, the city of the gods, etc.). Quoting William Jackson: «*The ideal form gracefully artificed suggests the infinite rising levels of existence and consciousness, expanding sizes rising toward transcendence above, and at the same time housing the sacred deep within. The gated enclosures-within-enclosures enshrine the inner sanctum, which for Hindus holds an external likeness of the inmost depths of divine mystery.* » (Jackson, 1999*).

In particular, Md Rian et al. (2007) affirmed: «*The underlying relationship between Hindu cosmology and fractal theory is manifested in Hindu temple where fractal geometry acts as the language*» (Md Rian et al., 2007, p. 4093). Figure 7 represents an example of Hindu temples which evidence a fractal organization.

Figure 7. Hindu temples which show a fractal organization.

Oriental Architecture

There are other examples of unconscious self-similarity in Oriental architecture (Rawson, 1990). Figure 8a illustrates the *Kaiyuan Si Pagoda*, Chinese architecture (Song Dynasty, 1228 – 1250, Quanzhou, Fuqian). Observing Figure 8b, which represents *Kaiyuan Si Pagoda*'s plan, we can note the self similarity in the octagonal shape which is repeated four times. We can interpret it as a kind of self-similarity which is also present in *Castel del Monte* (Apulia, Italy) (Sala, 2002; Sala and Cappellato, 2003a). Octagonal shapes appear in other pagodas (Liu, 1989). We can also find the presence of fractal geometry and the self-similarity in the *Sacred Stupa* Pha That Luang - Vientane (Laos), where the basic shape is repeated in different scales (see Figure 9) and in the *Royal Palace* of Mandalay (Burma, in Figure 10). (Sala, 2000; Sala and Cappellato, 2003a).

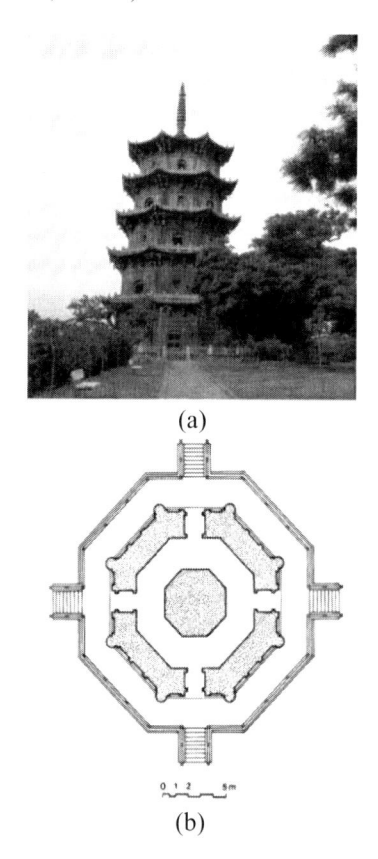

(a)

(b)

Figure 8. (a) *Kaiyuan Si Pagoda* (Song Dynasty, 1228 – 1250, Quanzhou, Fuqian) (b) *Kaiyuan Si Pagoda*'s plan which shows the self-similarity.

Figure 9. *Sacred Stupa* Pha That Luang - Vientane (Laos), the basic shape is repeated in different scales.

Figure 10. *Royal Palace* of Mandalay (Burma) evidences a fractal organization.

Modern Architecture

The conscious building's self-similarity is a recent discovery by the twentieth century architects as a result of a specific and conscious act of design. For example, Frank Lloyd Wright (1867-1959), in his *Palmer house* (Ann Arbor, Michigan, 1950-1951) used some self-similar equilateral triangles in the plan. He realized a kind of "nesting" of fractal forms that can be observed in two different points in the Palmer house: the entry way and the fireplace. At these places one encounters not only actual triangles but also implied (truncated) triangles. At the entrance there are not only the triangles composing the ceramic ornament, there is also triangular light fixture atop of triangular pier. The fireplace hearth is a triangular cavity enclosed between triangular piers. The concrete slab in which the grate rests is a triangle. The hassocks are truncated triangles (Eaton, 1998).

Remembering the definition of the fractal as « *a geometrical figure in which an identical motif repeats itself on an ever diminishing scale*», the *Palmer house* is an excellent instance of this concept. Other Wright's example of fractal architecture is the *Marin County Civic Center*, San Rafael (1957) where the self similarity is present in the external arches. Figure 11 shows Wright's *Marin County Civic Center* and Figure 12 illustrates a Roman aqueduct, the analogy in the shape is interesting.

Few people know that in 1908 the Catalan architect Antoni Gaudí (1852-1926) imagined a skyscraper for New York City. The building was drawn sometime between 1908 and 1911 having been ordered by an unknown American businessmen wanting a big hotel for New York. Unfortunately, the project was never realised; it got lost within the time and fell into oblivion. Only some original sketches survived as well as some drawings by sculptor Llorenç Matamala i Pinyol, friend and collaborator of Gaudí. The building, that Gaudí traced in five minuscule sketches on card paper, was an enormous construction that would have been the biggest of New York City at the time: 360 meters in height, something less than the *Empire State*, built in the 1931, and sixty meters less than the remembered *Twin Towers*. The shape of this rugged tower, in Figure 13, would evoke some reminiscences of his temple of the *Sagrada Familia* (begun in 1884, "last great sanctuary of Christendom", located in Barcelona, Spain). Gaudí's skyscraper is similar to the Hindu temple, shown in Figure 14, and this highlights that Catalan architect perceived the influence of the fractals in architecture (Sala, 2003).

Kazimir Malevich (1878 – 1935) was an important figure in Russian and Soviet art and architecture in the early of twentieth century. In the beginning of his artistic career he does «*not concerned with nature or analyzing visual impressions, but with man and his relation to the cosmos*» (Gray, 1962, p. 145). He designed some interesting examples of architectural projects, creating 3D models of buildings, that evidenced a fractal organization (as shown in Figure 15). The largest component of a building was repeated in a set of smaller and smaller copies characterized by an approximate 1/f relation.

Figure 11. Wright's *Marin County Civic Center* (1957, San Rafael, USA).

Figure 12. Roman aqueduct (20 B.C, Nimes, France).

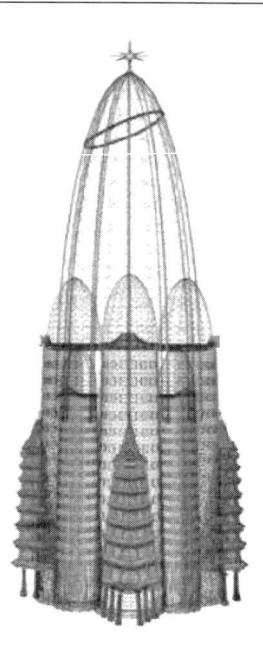

Figure 13. Gaudí's skyscraper (1908).

Figure 14. Hindu Temple.

Figure 15. Malevich's 3D model evidences a component which is repeated in smaller and smaller copies governed by an approximate 1/f relation.

2.1.3. Iterated Function System (IFS) in Architecture

Iterated Function System (IFS) is another fractal that can be applied in the architecture. Barnsley (1993, p. 80) defined the Iterated Function System as follow: «*A (hyperbolic) iterated function system consists of a complete metric space (X, d) together with a finite set of contraction mappings* $w_n: X \to X$ *with respective contractivity factor* s_n, *for* $n = 1, 2, ..., N$. *The abbreviation "IFS" is used for "iterated function system". The notation for the IFS just announced is* { X, w_n, $n = 1, 2, ..., N$ } *and its contractivity factor is* $s = max$ { $s_n : n = 1, 2, ..., N$ }».

Barnsley put the word "hyperbolic "in parentheses because it is sometimes dropped in practice. He also defined the following theorem (Barnsley, 1993, p. 81): "Let { X, w_n, $n = 1, 2, ..., N$ } be a hyperbolic iterated function system with contractivity factor s. Then the transformation $W : H(X) \to H(X)$ defined by:

$$W(B) = \cup_{n=1}^{n} w_n(B)$$

(3)

For all $B \in$ H($\mathbf{X}$), is a contraction mapping on the complete metric space (H($\mathbf{X}$), h(d)) with contractivity factor s. That is:

$$H(W(B), W(C)) \leq s \cdot h(B,C) \tag{4}$$

for all B, C $\in$ H($\mathbf{X}$). Its unique fixed point, A $\in$ H($\mathbf{X}$), obeys

$$A = W(A) = \cup_{n=1}^{n} w_{n}(A) \tag{5}$$

and is given by $A = \lim_{n \to \infty} W^{on}$ *(B)* for any B $\in$ H($\mathbf{X}$)."

The fixed point A $\in$ H($\mathbf{X}$), described in the theorem by Barnsley is called the "attractor of the IFS" or "invariant set".

Bogomolny (1998) affirms that two problems arise (Bogomolny, 1998). One is to determine the fixed point of a given IFS, and it is solved by what is known as the "*deterministic* algorithm".

The second problem is the inverse of the first: for a given set $A \in$ H($\mathbf{X}$), find an iterated function system that has A as its fixed point (Bogomolny, 1998). This is solved approximately by the Collage Theorem (Barnsley, 1993, p. 94). The Collage Theorem states: "Let ($\mathbf{X}$, d), be a complete metric space. Let $L \in$ H($\mathbf{X}$) be given, and let $\varepsilon \geq$ o be given. Choose an IFS (or IFS with condensation) $\{\mathbf{X}, (w_n), w_1, w_2,\ldots, w_n\}$ with contractivity factor $0 \leq s \leq 1$, so that

$$h(L, \cup_{n=1}^{n} w_{n}(L)) \leq \varepsilon \tag{6}$$
$$\scriptstyle (n=0)$$

where h(d) is the Hausdorff metric. Then

$$h(L, A) \leq \frac{\varepsilon}{1-s} \tag{7}$$

where A is the attractor of the IFS. Equivalently,

$$h(L, A) \leq (1 - s)^{-1} h(L, \cup_{\substack{n=1 \\ (n=0)}} w_n(L)) \tag{8}$$

for all $L \in H(X)$."

The Collage Theorem describes how to find an Iterated Function System whose attractor is "close to" a given set, one must endeavour to find a set of transformations such that the union, or collage, of the images of the given set under transformations is near to the given set. The IFS create the connection between the true mathematical fractals and the Nature. The next sections describe some applications of the IFS in the Gothic and in the modern architecture.

Gothic Architecture

The Gothic is a style developed in northern France that spread throughout Europe between the twelfth and sixteenth centuries. The adjective "Gothic" was first used during the later Renaissance by the Italian artist Giorgio Vasari (1511-1574) like a depreciative term. He wrote: «*Then arose new architects who after the manner of their barbarous nations erected buildings in that style which we call Gothic*».

Fulcanelli, the twentieth century most enigmatic alchemist, gave another explication of the term Gothic, which is connected to the language of the alchemy. He wrote: «*Some sagacious authors, observing the analogy which exists between Gothic and Goethic, have thought that there had to be a narrow relationship between Gothic Art and Goethic Art or magic. Art gotique is the deformation, for us, of word argotique whose homophony is perfect. Therefore the cathedral is a masterpiece of "d'art goth" o "d'argot".(Alcuni autori perspicaci, e non superficiali, colpiti dalla similitudine che esiste tra gotico e goetico, hanno pensato che ci dovesse essere uno stretto rapporto tra Arte gotica e Arte goetica o magica. Per noi art gotique non è altro che una deformazione ortografica della parola argotique la cui omofonia è perfetta. La cattedrale è quindi un capolavoro d'art goth o d'argot).*» (Fulcanelli, 2000, p. 46).

Some fractal components are present in the Gothic churches; an example is shown in Figure 16a which reproduces the facade of the *Reims' Cathedral* (1210-1241, Reims, France). The white arrows point out the fractal components (Sala and Cappellato, 2004, p. 86). The self-similarity is also present inside the *Notre Dame* (1163-1250, Paris, France), as shown in Figure 16b (Sala, 2005)

Gothic architecture can be observed using the Iterative Function System. The method is similar to the Wright's approach (Wright, 1996). He dissected a fern in to similar part, and he marked some triangles on these parts which are similar to the whole, as shown in Figure 17a. An affine maps was determined by how they map a single triangle to another triangle. This allowed Wright to convert out dissection of the fern into four affine maps. Figure 17b shows the original four parts together with a triangle corresponding to the whole fern, it is drawn in bold lines.

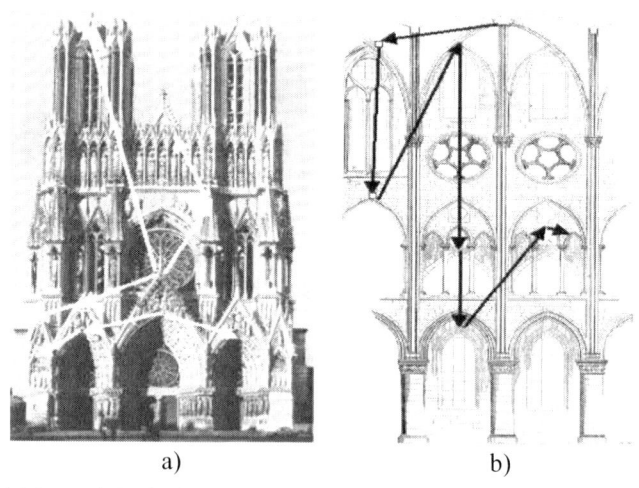

a)   b)

Figure 16. (a) *Reims' Cathedral* (1210-1241, Reims, France) and *Notre Dame* (1163-1250, Paris, France) b) show a fractal organization in their facade.

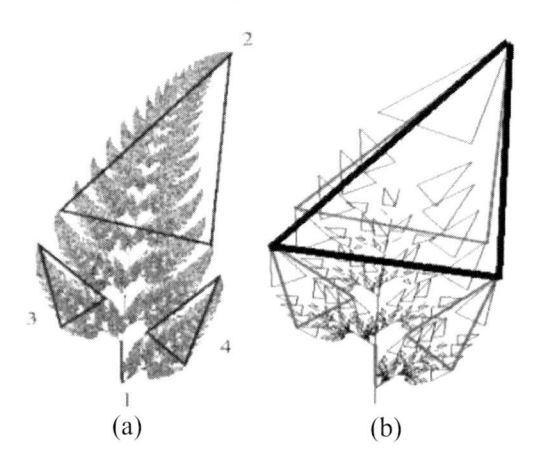

(a)   (b)

Figure 17. (a) Dissection of a fern into similar parts (b) mapping triangles for the fern. (Wright, 1996).

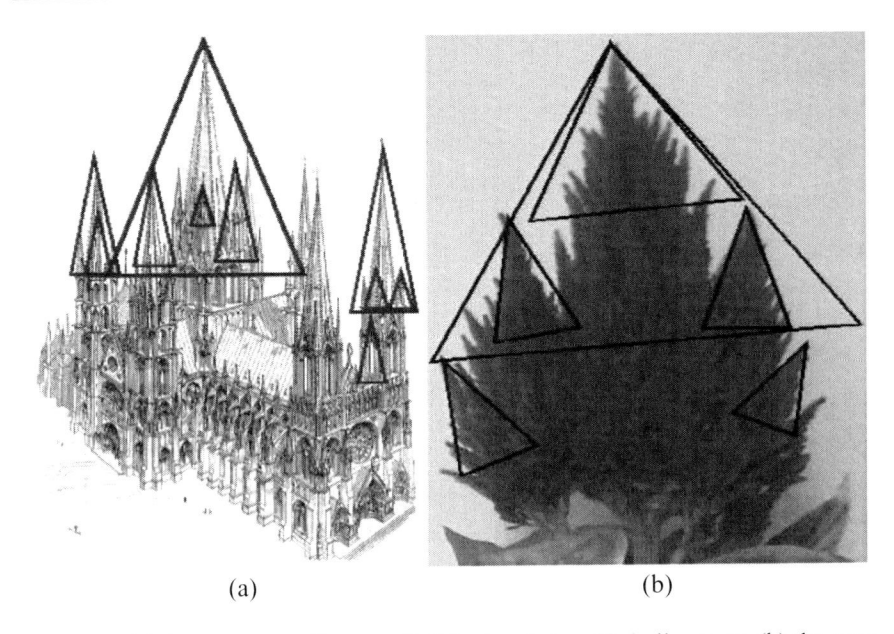

(a)   (b)

Figure 18. (a) An attempt to dissect a Gothic church in self-similar parts, (b) the same approach applied to a flower (Celosia plumosa).

Figure 18a illustrates an attempt to find a IFS which could generate the ideal *Gothic Church* conceived by Eugène-Emmanuel Viollet-le-Duc (1814-1879). In Figure 18b the same approach is applied to a flower (Celosia plumosa) which is manifestly fractal-like (Sala, 2005).

Modern Architecture

The Polish-Israeli architect Zvi Hecker, in his project of the *Heinz-Galinski Schule* (1993-1995, Berlin, Germany), recalls Iterated Function Systems. Hecker used spiral sunflower geometry (anticlockwise) plus concentric curves, self-similar curves and fish-shapes (see Figures 19a, and 19b) (Jencks, 1997). This institution, built in the suburbs of Charlottenberg for more than four hundred pupils, is the first Jewish school to be constructed in Germany for sixty years. The *Heinz-Galinski Schule* is a city inside a city, quoting Hecker: «*The school is a city within a city. Its streets meet at squares and the squares become courtyards. The walls of the schoolhouse also build walkways, passages, and cul de sacs. The outside of the school is also the inside of the city, because the school is the city.*» (Kiser, 2000). The *Heinz-Galinski Schule* creates a landform out of explicit metaphors. Mountain stairways, fish-shaped rooms and snake corridors (a snake path is shown in

Figure 19c), are pulled together with an overall sunflower geometry. The *Heinz-Galinski Schule* is playing an ethnic role in Berlin which is similar to Libeskind's *Jewish Extension* to the Berlin Museum: it must fit in and yet be unmistakably other. Hecker has underlined the paradox of *«a wild project»* that has *«very precise mathematical construction...Above all there is its cosmic relationship of spiral orbits, intersecting one another along precise mathematical trajectories.»* (Jencks, 1997, p. 25)

John Hejduk affirmed: *«Zvi Hecker's Jewish Community School in Berlin must be considered one of the major works in our time for its thought provoking energy that makes us think deep about many things related to life and architecture, not least about the meaning of knowledge, expulsion, place, and death.»* (Kiser, 2000)

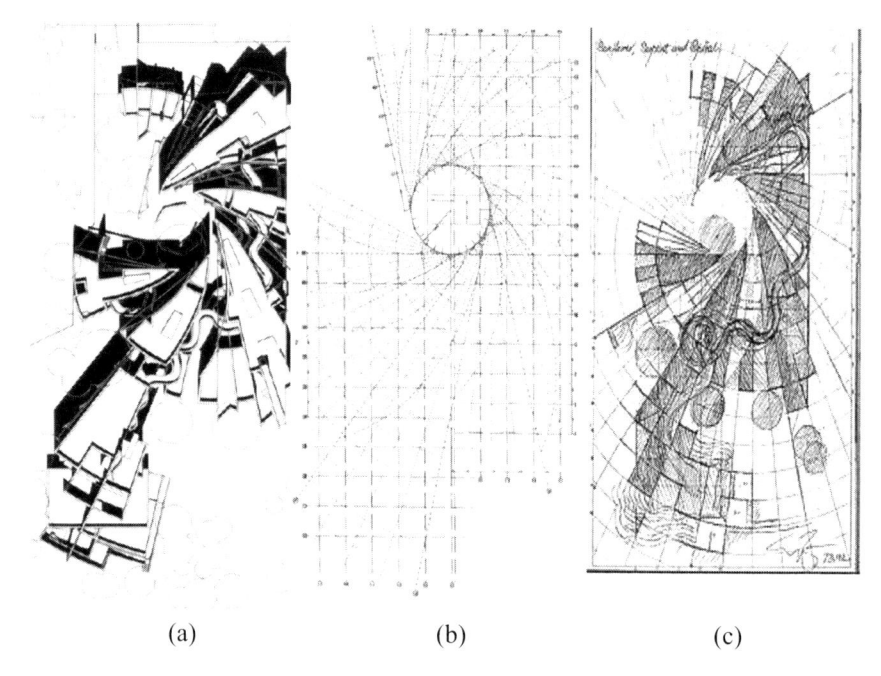

(a)   (b)   (c)

Figure 19. *Heinz-Galinski Schule* (1993- 1995, Berlin, Germany), designed by Hecker.

2.2. Fractals in the Large Scale Analysis

The applications of fractals in the large scale analysis are oriented to study the human settlements, the urban growth and the urban development (Batty, 1991; Eglash, 1999; Marsault, 2006; Saleri, 2002, 2004, 2006b; Thomas et al.,

2008). In particular, the self-similarity could be an useful property to understand the organization of the human settlements in a territory, in connection to the different cultures.

Eglash, in his book entitled *African Fractals* (1999), described interesting fractal examples in the African architecture arts, and design. Eglash focused his attention on the fact that the African architecture reflects the social structure, and the religious structure of a settlement.

In the architectural examples shown in the book, the author presented the fractal components as a direct consequence of a few structural and organizational features of a settlement. Eglash affirmed that the European expert in urban planning consider African settlements as big villages and not of true and real towns. Because, the Euclidean geometry has been not used for tracing the roads and for the urban organization, but complex forms which remember fractal shapes. Quoting Eglash: «*Thus fractal architecture was used as colonial proof of primitivism....During the development of colonial cities, the chaos of African architecture was used as both symbol and symptom of European fears over social chaos*»(Eglash, 1999, p. 196).

An interesting instance of social hierarchy inside a settlement is provided by Ba-ila (Southern of Zambia), where each house has a ring form, with the fence for the cattle, and has a main entry, and an exit. The whole settlement reproduces in every part the form of the main ring, and all rings are directly proportional to the social condition of the owners.

Figures 20a and 20b respectively show an aerial vision of a settlement Ba-ila and its schematization which denotes the self-similarity. In this example, the self-similarity, applied to the house structures Ba-ila, becomes a way to present the "social status" of a family inside its community.

The geographer Michael Batty suggested that the fractal geometry can describe the urban growth: «*The morphology of cities bears an uncanny resemblance to those dendritic clusters of particles which have been recently simulated as fractal growth processes.*»(Batty, 1991)

In the book *Fractal Cities* (1994), Batty and Longley introduced a fractal generation of the cities using the cellular automata models, and they have interpreted the results in a context of self-organization. Their work was a pioneering study of the development and of the use of fractal geometry for understanding and planning the physical form of cities. They showed how this kind of geometry permits to simulate the urban growth of the cities using computer graphics tools.

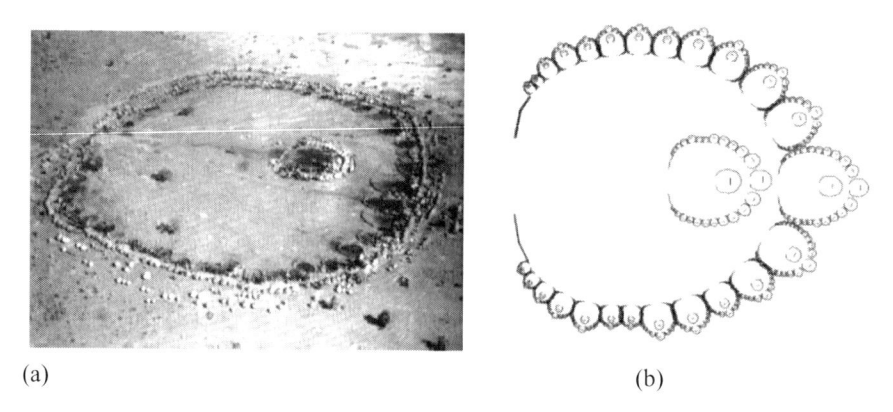

(a) (b)

Figure 20. Aerial photo of Ba-ila settlement before 1944 a) Fractal generation of Ba-ila simulation b) (Eglah, 1999, p. 27).

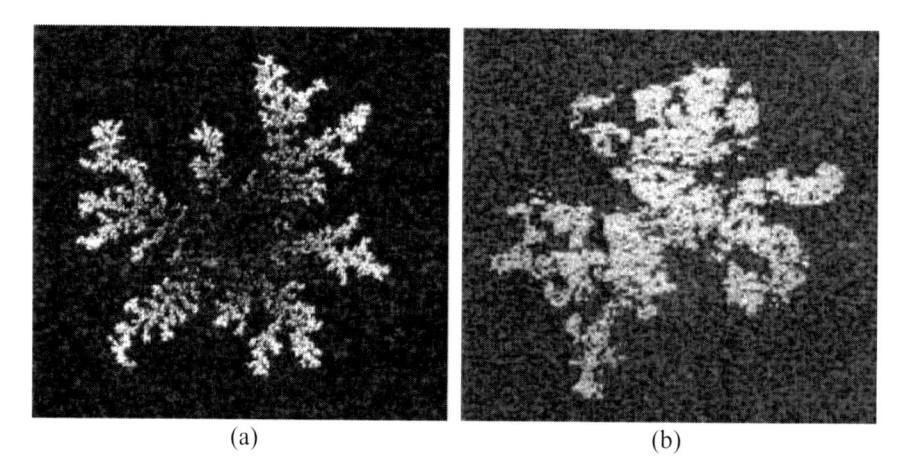

(a) (b)

Figure 21. (a) A typical DLA simulation. (b) Urban development in Taunton at 1981 digitised from 1:10 000 Ordnance Survey map.

The study of urban patterns using fractal investigation is based on two aspects: the direct study of spatial organisation, and the formalization of self-generating geometrical structures. The growth of urban models is described by time-based spatial simulators or by simple static generators. Spatial simulators are usually based on simple "life-game" (cellular automata) devices or by Diffusion Limited Aggregation (DLA). DLA is the process where particles, that follow a random walk due to Brownian motion, cluster together to form aggregates of such particles. Witten and Sander proposed DLA model in 1981, and it is applicable to aggregation in the systems where the diffusion is the primary means of transport (Witten and Sander, 1981). The term "Diffusion" was used because the particles forming the structure wander around randomly before attaching themselves ("Aggregating") to the structure. "Diffusion-limited" because the particles are considered to be in low concentrations so they do not come in contact with each other and the structure grows one particle at a time rather then by chunks of particles. Figure 21a shows a DLA simulation, Figure 21b illustrates Urban development in Taunton at 1981 generated by computer (Batty and Longley, 1994).

2.2.1. L-System in Urban Generation

An L-system is a parametric rewriting system operating on branching structures represented as bracketed strings of modules. It was introduced by the Hungarian biologist Aristed Lindenmayer (1925-1989) in 1968. He proposed the formalism of L-systems which concerned in a mathematical theory of plant development (Lindenmayer, 1968). In the context of L-systems, the term "module" represents any discrete constructional unit which is repeated as the plant develops, for example a branch or a flower L-systems provided a formal description of the development of such simple multicellular organisms, and they also illustrated the neighbourhood relationships between plant cells.

L-systems can help to realize the connections between cognitive design processes and the composition of software code which generates architectural forms autonomously. Saleri (2002, 2006a, 2008) studied different automatic generative methods able to produce architectural and urban 3D-models. In the first approach they established an alphabet: 0,1,[,] (Saleri, 2002).

In the example, 0 and 1 occurrences "produce geometry "while [and] provide a simple affine transformation (rotation and/or translation). Simple substitution rules was applied to alphabetic elements:

0 : 1[0]1[0]0 **1** : 11 **[** : **[**] :]

Applying in recursively way these substitution rules to an initial sprout (applied from the top to the rule of letter "0") we have:

11 [1[0]1[0]0] 11 [1[0]1[0]0]1[0]1[0]0

Two "generations" or recursive steps are:

11 11 11 11 [11 11 [11 [1[0]1[0]0] 11 [1[0]1[0]0] 1[0]1[0]0] 1111 [11 [1[0]1[0]0] 11 [1[0]1[0]0]1[0]1[0]0] 11[1[0]1[0]0] 11 [1[0]1[0]0]

1[0]1[0]0] 11 11 11 11 [11 11 [11 [1[0]1[0]0] 11 [1[0]1[0]0] 1[0]1[0]0]
1111 [11 [1[0]1[0]0] 11 [1[0]1[0]0]1[0]1[0]0] 11 [1[0]1[0]0] 11 [
1[0]1[0]0] 1[0]1[0]0] 11 11 [11 [1[0]1[0]0] 11 [1[0]1[0]0] 1[0]1[0]0] 11
11 [11 [1[0]1[0]0] 11 [1[0]1[0]0] 1[0]1[0]0] 11[1[0]1[0]0] 11 [1[0]1[0]0
] 1[0]1[0]0

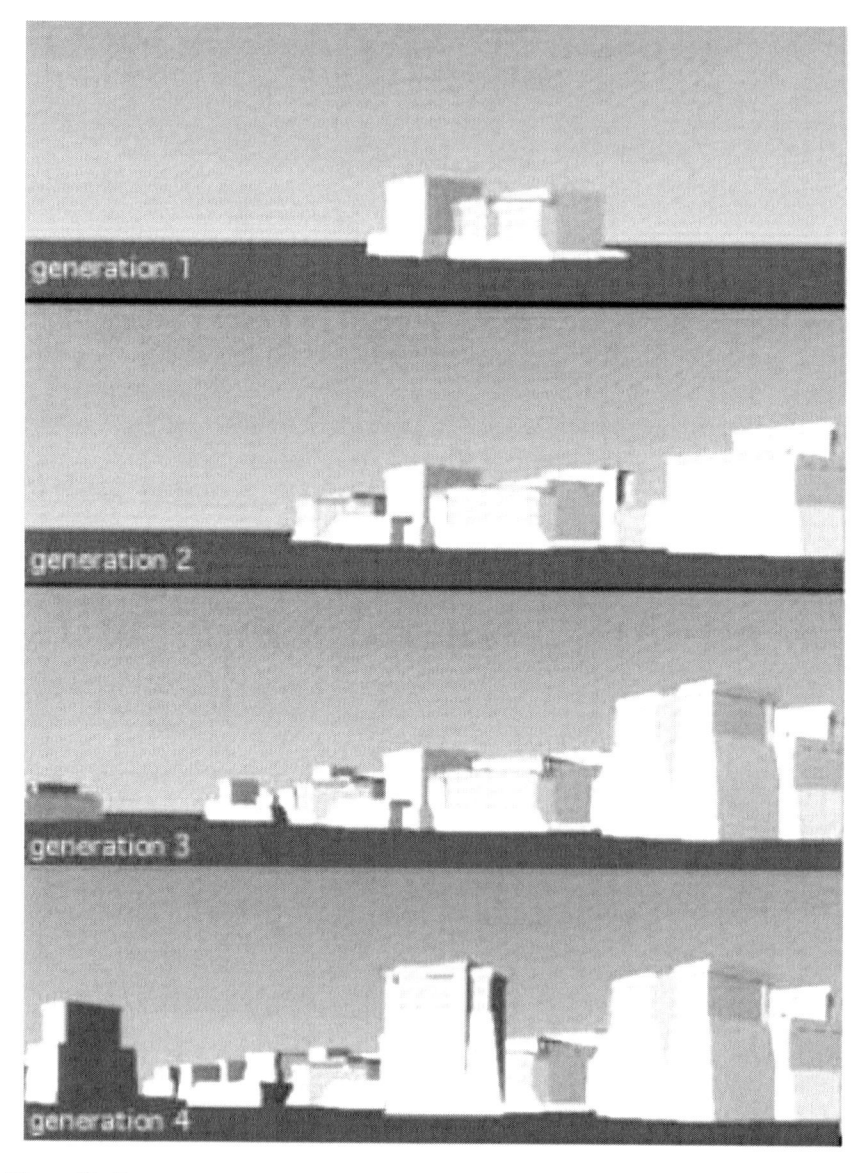

Figure 22. Four-steps generated VRML model, based on two architectural primitives.

They replaced the brackets by specific 3D operations, typically affine transformations, such as rotations or translations, and the "0" and "1" occurrences by 3D pre-defined objects. Iterative instructions written in VRML (Virtual Reality Modelling Language) generated a model. Figure 22 shows the results of four-steps generated VRML model, realized by two architectural primitives.

Saleri presented a research project that consisted in the application of automatic generative methods in design processes (Saleri, 2008). This approach consisted in an integrated architectural and urban semiautomatic model-generation pipe, emerging from early research tasks about automatic generation of urban and architectural 2D and 3D patterns proposed by him in 2004. The goal in this research task was to rapidly produce "plausible" urban environments, using existing data, such as digital maps, and aerial photographs with a high-level of detail 16 or 50 cm resolution.

Early stages of this project produced interesting results, combining complementary modeling techniques, according to demanded LOD (Level of Detail). For instance, the use hybrid image-based modeling for relevant architectural objects, demanding high-level recognisability, for close-up views and close detail identification. If not specified, the model generation follows a generic approach. The semi-automated process involved in rapid 3D-modeling for generic surrounding architecture (architectural sceneries) links two semi-automated generative processes considering separately 3D elevation and facade generation: (i) the 3D elevation step, and (ii) the facade generation step (Saleri, 2008).

3. HYPER ARCHITECTURE AND BEYOND...

In a period which is dominated by the electronic media, by the networks and by the new Information Communication Technologies the perception of space has had an intensive transformation. Today, traditional architectural space is changing and the buildings are being rendered transparent, fleeting and intangible, enhanced by virtual potential.

Several theories have been developed in these decades: Liquid Architecture, Transarchitecture, Hyperarchitecture, Hypersurfaces, and Virtual Architecture.

Marcos Novak, architect, artist and musician was one of the first people to show to the architects the way into new realm of cyberspace. Novak affirmed:

«Cyberspace alters the ways in which architecture is conceived and perceived beyond computer-aided design (CAD), design computing (DC), or the development of new formal means of describing , generating, and transforming architectural form, encodes architectural knowledge in a way that indicates that our conception of architecture is becoming increasingly musical, that architecture is spatialized music [...] This, of course, means that any information, any data, can become architectonic and habitable, and that cyberspace and cyberspace architecture are one and the same. A radical transformation of our conception of architecture and public domain that is implied by cyberspace. The notions of city, square, temple, institution, home, infrastructure are permanently extended.» (Novak, 1991, p. 1)

Novak, in his seminal 1991 essay *Liquid Architecture of Cyberspace*, equated the notions of cyberspace and virtuality as a metaphor of liquid. He coined the terms "Liquid Architecture" as a clear example of dematerialized architecture (an example is shown in Figure 23).

«I use the term liquid to mean animistic, animated, metamorphic, as well a crossing categorial boundaries, applying the cognitively supercharged operations of poetic thinking. Cyberspace is liquid. Liquid cyberspace, liquid architecture, liquid cities.[...] If we described liquid architecture as a symphony in space, this description would still fall short of the promise. A symphony, though it varies within its duration, is still a fixed object and can be repeated. At its fullest expression a liquid architecture is more than that. It is a symphony of space, but a symphony that never repeats and continues to develop. If architecture is an extension of our bodies, shelter and actor for the fragile self, a liquid architecture is that self in the act of becoming its own changing shelter. Like us, it has an identity; but this identity is only revealed fully during the course of its lifetime.» (Novak, 1991, p. 1)

Novak is also a pioneer of another theory: transarchitecture, where he uses mathematical algorithms to build virtual, hybrid and clever spaces. The term "transarchitecture" was coined to refer one possible way to begin thinking about the construction of hybrid, physical and virtual, spaces. Figure 24 shows an example of transarchitecture.

Transarchitecture connected architecture and new technologies, in particular Virtual Reality. It described an architecture transformation towards the break of the opposition of physique and virtual and the proposal of a continuum which can conduce from a physical architecture to an architecture technologically strengthened to an architecture of cyberspace.

About transarchitecture Neil Spiller, a pioneer into impact of advanced technology on architectural design and practice, affirmed:

«We conceive algorithmically (morphogenesis); we model numerically (rapid prototyping); we build robotically (new tectonics); we inhabit interactively (intelligent space); we telecommunicate instantly (pantopicon); we are informed immersively (liquid architectures); we socialise nonlocally (nonlocal public domain); we evert virtuality (transarchitectures).»(Spiller, 2004)

Novak also observed the biological world and the organic form in the design of his works, but referring above all to the processes of organic generation of the form. He does not only simulate some generative and organic process to realize new physical or virtual structures, but he wishes to produce new vital and real kinds: the "Allo", suffix which shows the generation of the stranger, of "other of another kind" (De Feo, 2008). Figure 25 shows *Allomorphic forms* (2002) conceived by Novak.

Figure 23. Example of liquid architecture and information forms which are floating through cyberspace by Novak.

Figure 24. Transarchitecture by Novak.

Figure 25. *Allomorphic forms* (2002) conceived by Novak.

Hyperarchitecture (or Hyper-Architecture) is an architecture characteristic of the information age. It postulates a space which has definitively abandoned the here and the time.

Prestinenza Puglisi for presenting Hyperarchitecture wrote:

«*In the architecture of this last decade, several very important discoveries have been made regarding the figurative arts. Gehry in his last period owes quite a lot to Boccioni and his concept of trajectory, to that force of going beyond the plastic quality of the isolated object toward an atmospheric vibration. Peter Eisenman has adopted more than technique from the vibration of Duchamp and Balla. The dripping technique of Pollock is touched upon in various types of research into new forms of landscape and construction of nature...We know that artists have a spatiality which transmigrates into architecture. But the fluid, liquid submarine, metaphorical, symbolic and interconnected spatiality of Kandinsky (and Mirò and Klee) is, without information technology, impossible to conceptualize in architecture. With information technology, on the other hand, this becomes almost vaguebly intuitable. This is perhaps what we are attempting to do with this strange word: Hyper-Architecture ...*».(Prestinenza Puglisi, 1999, pp.85-86)

Hyperarchitecture not only want to be a narrative and metaphorical architecture but also interactive, that can have structures, spaces and situations navigable and modifiable as a hypertext. The structure of the hypertext is the key of the connections between the information, since it allows to look for what which one needs through ways (Prestinenza Puglisi, 1998).

Three phenomena can characterize Hyperarchitecture: a) the loss of the traditional concept of place, b) the end of the distinction between animated and lifeless, c) the reflection on the shape-information-relationship triad. Hyperarchitecture is moved by a new intuition on the virtual space configurations, looking for an integration between information, technology and users; where the empiric place coexists with the virtual realities of other spaces which break and superimpose themselves in the same way in which various windows are overlapped on the computer.

Other interesting evolutions in the architectural theories are connected to the concept of Hypersurface. In mathematics it is a surface in hyperspace, but in architectural field the hypersurface is a new concept that promotes broader interfaces and interactivity between cyberspace and the build environment. Stephen Perrella extends the mathematical term and here the hypersurfaces are rethought to render a more complex notion of space-time-information.

Perrella affirmed (1995):

«A hypersurface is a new theory of liquid-embodied architecture to displace the nostalgia and re-realization being carried into the spatial conceptions of new-media technology. We shouldn't think cyberspace with conventional assumptions. Hypersurface delimits reductions assumed in biases prevalent in disciplinary categorisations. Epistemological thought hasn't produced what it promised prior to its entry into cyberspace; there are only further degradations to come...Hypersurface architecture is a way of thinking about architecture that does not assume real/irreal, material/immaterial dichotomies. It is to consider an architecture prior to those assumptions, that entails a condition also prior to the assumption of a split between body-subject/building. To think this architecture is not an act of construction or deconstruction but a nearly self-generating between-state. The generation of it occurs in an interplay and interaction between the delimited forces, energies and desire/life in substance (Deleuze) and language (Derrida). The architectonic translation of surface is structure/substrate».

He also confirmed (1997):

«As a verb hypersurface considers ways in which the realm of representation (read images) and the realm of instrumentality (read forms) are respectively becoming deconstructed and deterritorialized into new image-forms of intensity. Hypersurfaces are an interweaving and subsequent unlocking of culturally instituted dualities. Hypersurface theory is not a subjective invention in contrast to what seems an unending foray of "isms" attempting to explain postmodern culture, (for instance in the efforts of Charles Jencks)». (Perrella, 1997, p. 7)

Figure 26 shows The *Moebius house* (1997-1998) a study conceived by Stephen Perrella and Rebecca Carpenter.

Other theory that connects virtual reality, simulation and mapped data is Virtual Architecture. It has been introduced by Hani Rashid (2003) as:

«The virtual architecture is an evolving discipline which results from the convergence of mapped data and of simulation, production of the digital form, architecture of the information, buildings and theory of the virtual reality [...]. The conventional architectures base themselves on the stay and geometrical certainty, while virtual architecture uses digital technologies to widen the real events, the time and the space." (L'architettura virtuale è una disciplina in evoluzione che risulta dalla convergenza di dati mappati e simulazione, produzione della forma digitale, architettura dell'informazione, costruzioni e teoria della realtà virtuale [...] Le architetture convenzionali tendono a basarsi sulla permanenza e certezza geometrica, mentre l'architettura virtuale utilizza tecnologie digitali per allargare gli eventi reali, il tempo e lo spazio)». (Rashid, 2003, p. 169).

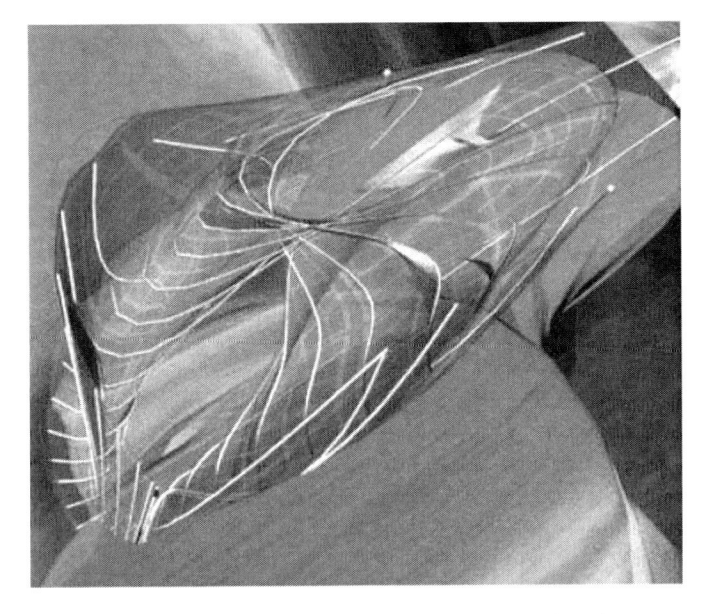

Figure 26. *Moebius House* (1997-1998) trasversal NURB animations, conceived by Stephen Perrella and Rebecca Carpenter.

To understand the meanings and the objectives of the virtual architecture is useful distinguish three different disciplinary aspects (Musella, 2008). The first (instrumental aspect) is tight connected to the digital representation.

The second aspect considers the virtual architecture in the electronic drawing as different morphological suggestion, where the digital one produces the matrix with which search for new spatiality; a world in which test buildings to the limit of the physical laws. The computer becomes structuring part of the creative process suggesting transparency, fluidity, dynamism, plasticity, and removing every perceptive reference of the bearing elements.

The third aspect, the most interesting, is the experimental one. Virtual architecture is seen in the spatiality of the digital medium, completely autonomous from the reality (Musella, 2008).

CONCLUSION

Fractal geometry, new communication and information technologies and their connections with the complexity can help to introduce new paradigms in architecture (Batty and Longley, 1994; Bovil, 1995; Perrella, 1995, 1997, 1998; Frankhauser, 1997; Prestinenza Puglisi, 1998, 1999; Salingaros, 1999,

2000, 2001; Gregory, 2003; Sala and Cappellato, 2003a, Capo, 2004; Sala and Cappellato, 2004; Semboloni, 2006; Thomas et al. 2008). This paper introduces an approach that observes the architecture using a fractal point of view applied in small and in large scale analyses. For example, the property of the self-similarity in the Gothic and the Baroque styles has been chosen for an aesthetic sense (the architects of that periods did not know the fractal geometry, because it is a recent discovery). Thus, it is possible to refer as an "unintentional" use of the fractal geometry. The Iterated Function System applied in the Gothic cathedrals could help us to understand the generative processes of these complex buildings (Sala, 2005).

In modern architecture the self-similarity appears in intentional way as the result of a specific and conscious act of design (Bovil, 1995; Eaton, 1998; Sala, 2000, 2002; Sala and Cappellato, 2004).

Human and computer interaction within the design realm exists and fractal geometry can help to realize the connections between cognitive design processes and the composition of software code which generates architectural forms autonomously.

For example, L-systems are studied for creating some automatic generative methods able to produce architectural and urban 3D-models. Recent studies reveal that the IFS could help to create a new pseudo urban models based on fractal algorithms (Saleri, 2006b). Thus, it could be possible to encode simplified 2D½ city models using an IFS compression technique.

The age of information has modified the architecture introducing virtual world and new theories on architecture.

Hypersurface theory promotes increased accessibility to the Internet, initiates new ideas regarding architectural ornament and instigates new explorations of architectural surfaces and materials, but new architecture is not bound any more to the production in series and the objectivity but assumes a symbolic language, figurative, therefore subjective where predominate the information and communication and the individual aesthetic research.

New information and communication technologies also introduce the concepts of Virtual and of Virtual Reality (VR).

The virtual one can be seen as a differentiation process.

Quoting Peter Eisenman (1997):

«The virtual constitutes the entity: the virtualities inherent in a being, its problems, the node of tensions, constraints and projects that brings it to life, the questions that give it movement, are an essential part of its determination.» (Eisenman, 1997)

Some architectural theorists are looking at virtual reality as such an inhabitable alternative reality. Whyte (2002) affirms:

«They describe objects in interactive, spatial, real-time media as though they existed in a new form of space, rather than in spatial representations and look at Novak (1996) terms the vitality of architecture after territory.» (Whyte, 2002, p. 46).

And beyond?
VR is also connected to the cyberspace.

William Gibson, in his cyber novel *Neuromancer* (1984), introduced cyberspace as:

«A consensual hallucination experienced daily by billions of legitimate operators, in every nation, by children being taught mathematical concepts... A graphical representation of data abstracted from the banks of every computer in the human system. Unthinkable complexity. Lines of light ranged in the non-space of the mind, clusters and constellations of data.» (Gibson, 1984, p. 51).

Novak (1996), recalling Gibson's cyberspace, affirmed:

«Cyberspace as a whole, and networked virtual environments in particular, allow us not only to theorise about potential architectures informed by the best of current thought, but to actually construct such spaces for human inhabitation in a completely new kind of public realm. This does not only imply a lack of constraint, but rather a substitution of one kind of rigour for another. When bricks become pixels, the tectonics of architecture become informational. City planning becomes data structure design, construction costs become computational costs, accessibility becomes transmissibility, proximity is measured in numbers of required links and available bandwidth. Everything changes, but architecture remains.» (Novak, 1996)

For Michael Benedikt (1992)

«Cyberspace is architecture; cyberspace has an architecture; and cyberspace contains architecture.»

Cyberspace can use fractal algorithms for generating new shapes and new buildings in its virtual worlds.

What is the future of the application of fractals in architecture on large scale?

Recent studies are testing different hypotheses. For example, micro-dynamic models of urban development (Semboloni, 2006). In other studies, fractal dimensions allow a synthetic description of the built environment of each commune; fractal dimensions can be used to evaluate the quality of the built environment of each commune. More precisely, that the multi-scale organisation of a built-up area reveals the good qualities of that environment from a functional (not an aesthetic) point of view (Thomas et al., 2008).

In other studies L-systems are applied for creating some automatic generative methods able to produce architectural and urban 3D-models (Saleri, 2008).

In the last years, the practice of architecture has changed in radically way. Hardware technologies and complex software have created a globally transferable design culture and new communities. Important key words for the future of the architecture will be: dynamic data interconnection, data processing models, interactivity, new spaces. Fractal geometry will help to understand the hidden order tight connected to the dynamic evolutions of the urban growth.

In all these fields the future could be completely re-thought.

REFERENCES

Barnsley, M.F. (1993). *Fractals everywhere*, Academic Press, Boston, 2nd edition.

Batty, M. (1991). *Cities as Fractals: Simulating Growth and Form*, A. J. Crilly, R. A. Earnshaw and H. Jones, *Fractals and Chaos*, Springer-Verlag, New York, pp. 41-69.

Batty, M., and Longley, P. (1994). *Fractal Cities*, London Academic Press, London.

Batty, M., and Longley, P. (1997).*The Fractal City,* Architectural Design, New Science = New Architecture Academy Group, London, n. 129, pp. 74 – 83.

Benedikt, M. (ed.) (1992). *Cyberspace: First Steps*, The MIT Press.

Bogomolny, A. (1998). *The Collage Theorem*. Retrieved September 15, 2005, from: http://www.cut-the-knot.org/ctk/ifs.shtml

Bovill, C. (1995).*Fractal Geometry in Architecture and Design*, Birkhäuser, Boston.

Burkle-Elizondo, G., and Sala, N. (2004). Complexity and Fractal Dimension in 26 Mesoamerican Pyramids, *Proceedings 7th Generative Art Conference*, Milano, Italy, vol. 1, pp. 206-214.

Burkle-Elizondo, G., Sala, N., and David Valdez-Cepeda, R. (2004). Geometric and Complex Analyses of Maya Architecture: Some Examples, in *Nexus V: Architecture and Mathematics*, Williams, K., and Delgado Cepeda F. (eds.), Fucecchio (Florence): Kim Williams Books, pp. 57-68.

Capo, D., (2004). The Fractal Nature of the Architectural Orders, *Nexus Network Journal* vol. 6 no. 1, pp. 30-40.

De Feo, C. (2008). Marcos Novak: Il futuro è "Allo", *Digimag*, issue 33, http://wwww.digicult.it/digimag/article.asp?id=1133.

Donato, F., and Lucchi Basili, L. (1996). *L'ordine nascosto dell'organizzazione urbana* [The hidden order of the urban organization], Franco Angeli Editore, Milano.

Eaton, L.K. (1998). Fractal Geometry in the Late Work of Frank Llyod Wright: the Palmer House, Williams, K. (ed.), *Nexus II: Architecture and Mathematics*, Edizioni Dell'Erba, Fucecchio, pp. 23–38.

Eglash, R. (1999). *African Fractals: Modern Computing and Indigenous Design*, Rutgers University Press, Piscataway.

Eiseman, P. (1997). Interview, *Progetto*, n. 1.

Falconer, K. (1990). *Fractal Geometry in Architecture and Design*, Chicester: Wiley.

Frankhauser, P. (1994). *La Fractalité des Structures Urbaines* [The Fractality of Urban Structures], Collection Villes, Anthropos, Paris, France.

Frankhauser, P. (1997). *L'approche fractale: un nouvel outil de réflexion dans l'analyse spatiale des agglomérations urbaines* [The fractal approach: a new tool of reflection in the spatial analysis of urban agglomerations], Université de Franche-Comté, Besançon.

Fournier, A., Fussel, D., and Carpenter, L. (1982). Computer Rendering of Stochastic Models, *Communications of the ACM*, 25, pp. 371-384.

Fulcanelli, (2000). *Il mistero delle cattedrali e l'interpretazione esoterica dei simboli ermetici della Grande Opera* [The mystery of cathedrals and the esoteric interpretation of the Hermetic symbols of the Great Work], Edizioni Mediterranee, Roma.

Gibson, W. (1984). *Neuromancer*. Ace, New York.

Gray, C. (1962).*The Russian Experiment in Art 1863 – 1922,* Thames and Hudson, London.

Gregory, P. (2003), *Territori della complessità* [Territories of complexity], ed. testo&immagine, Torino.

112   Nicoletta Sala

He, X., Wang, H., Wu, Q., Hintz, T., and Hur, N. (2006). Fractal Image Compression on Spiral Architecture, *International Conference on Computer Graphics, Imaging and Visualisation (CGIV'06)*, pp. 76-83.

Hersey, G. (1999).*The Monumental Impulse*, MIT Press.

Hersey, G. (2000). *Architecture and Geometry in the Age of the Baroque*, University of Chicago Press, Chicago.

Kiser, K. (2000). *Zvi Hecker The Heinz Galinski School Berlin, Germany.* Retrieved, 2 June, 2009 from: http://www.arcspace.com/architects/zvi_hecker/heinz_galinski/index.html

Jackson, W.J. (1999*, in the Internet site does not appear the date). *Hindu Temple Fractals*, Retrieved 10, June 2008, from: http://liberalarts.iupui .edu/~wijackso/tempfrac/

Jackson, W.J. (2004). *Heaven's Fractal Net: Retrieving Lost Visions in the Humanities*, Indiana University Press.

Jencks, C. (1997). *Landform Architecture Emergent in the Nineties*, Architectural Design, New Science = New Architecture, Academy Group, London, n. 129, pp. 15 – 31.

Leland,W.E., Taqqu, M.S., Willinger, W., and Wilson, D.V. (1993). On the Self-Similar Nature of Ethernet Traffic, *Proceedings of the ACM/SIGCOMM'93*, (pp. 183-193).San Francisco, CA.

Lindenmayer, A. (1968). Mathematical models for cellular interactions in development, parts I-II, *Journal of Theoretical Biology* 18: 280-315.

Liu, L. G. (1989).*Chinese Architecture,* Academy Press, London.

Mandelbrot, B.B. (1975), *Les objects fractals. Forme, Hasard et Dimension* [Fractal objects. Shape, Chance and Dimension], Flammarion, Paris.

Mandelbrot, B.B. (1982). *The fractal geometry of nature*. W. H. Freeman, San Francisco.

Marsault, X. (2006). Generation of Texture and Geometric Pseudo-Urban Models with the Aid of IFS. Sala, N. (ed.), *Chaos and Complexity in Arts and Architecture*, Nova Science, New York, pp. 147-160.

Md Rian, I., Park, J.-H., Uk Ahn, H., and Chang, D. (2007). Fractal geometry as the synthesis of Hindu cosmology in Kandariya Mahadev temple, Khajuraho, *Building and Environment*, Volume 42, Issue 12, December 2007, pp. 4093-4107.

Mitchell, G. (1988). *The Hindu Temple: An Introduction to Its Meaning and Forms*, University of Chicago Press, Chicago.

Mitchell, G. (2000).*Hindu Art and Architecture (World of Art),* Thames and Hudson, London.

Musella, G. (2008). *TranArchitecture' Theory*. Retrieved 10 June, 2009 from: http://transarchitecture.wordpress.com/

Nonnenmacher, T.F., Losa, G.A., Merlini, D., and Weibel, E.R. (eds.). (1994). *Fractal in Biology and Medicine*. Basel, Switzerland: Birkhauser.

Novak, M. (1991). Liquid Architecture in Cyberspace, Benedikt, M. (ed.) *Cyberspace: First Steps*, MIT Press, Cambridge, MA, p. 1.

Novak, M. (1996). *Transmitting architecture: the transphysical city*, available: http://www.ctheory.net/text_file.asp?pick=76.

Oppenheimer, P. (1986).Real time design and animation of fractal plants and trees, *Computer Graphics*, 20(4), pp. 55–64.

Ostwald, M.J. (2001). "Fractal Architecture": Late Twentieth Century Connections Between Architecture and Fractal Geometry, *Nexus Network Journal*, vol. 3, no. 1 (Winter 2001). Retrieved, 15 December, 2008, from: http://www.nexusjournal.com/Ostwald-Fractal.html

Perrella, S. (1995). http://www.mediamatic.nl/Doors/Doors2/Perrella/Perrella-Doors2-E.html

Perrella, S. (1998). Hypersurface Theory: *Architecture><Culture** Retrieved, 15 March, 2009, from: http://architettura.supereva.com/extended /19981201/index_en.htm.

Perrella S. (ed.) (1997). *Hypersurface Architecture*, Academy Editions, London

Prestinenza Puglisi, L. (1998). *HyperArchitettura*, n.38 coll. Universale di Architettura, ed. testo&immagine, Torino.

Prestinenza Puglisi, L. (1999). *HyperArchitettura: Spaces in Electronic Age*, Birkhäuser, Basel.

Prusinkiewicz, P. (1986). Graphical applications of L-systems, *Proceedings of Graphics Interface '86 - Vision Interface '86*, pp. 247–253.

Prusinkiewicz, P., and Lindenmayer, A. (1990). *The Algorithmic Beauty of Plants*. New York, US: Springer-Verlag. Retrieved September, 2006, from: http://algorithmicbotany.org/papers/abop/abop.pdf

Rashid, H. (2003). Architettura virtuale, Sacchi, L., and Unali, M.(eds.) *Architettura e cultura digitale* [Architecture and digital culture], Skirà, Milano, pp.169-178.

Rawson, P. (1990).*The Art of Southeast Asia*, Thames and Hudson, London.

Sala, N., and Cappellato, G. (2003a). *Viaggio matematico nell'arte e nell'architettura* [Mathematical journey in art and architecture], Franco Angeli Editore, Milano.

Sala, N., and Cappellato, G. (2003b). The generative approach of Botta's San Carlino. *Proceedings 6th Generative Art Conference,* Milano, Italy, pp. 328-337.

Sala, N., and Cappellato, G. (2004). *Architetture della Complessità. La geometria frattale tra arte, architettura e territorio* [Architectures of Complexity. Fractal geometry between art, architecture and territory], Franco Angeli Editore, Milano.

Sala, N. (2000). Fractal Models In Architecture: A Case Of Study, *Proceedings International Conference on "Mathematics for Living",* Amman, Jordan, November 18-23, pp. 266–272.

Sala, N. (2002). The presence of the Self- Similarity in Architecture: Some examples, Novak, M.M. (ed.), *Emergent Nature*, World Scientific, pp. 273–283.

Sala, N., (2003b) Fractal Geometry And Self-Similarity In Architecture: An Overview Across The Centuries, *Isama-Bridge 2003 Conference Proceedings,* Granada, Spagna, pp. 235-244.

Sala, N. (2005). Fractal components in the Gothic and in the Baroque Architecture, *Proceedings 8h Generative Art Conference,* Milano, Italy, vol. 1, available: http://www.generativeart.com/on/cic/papers2005/27. NicolettaSala2005.htm.

Sala, N. (ed.) (2006). *Chaos and Complexity in Arts and Architecture*, Nova Science, New York

Sala N., Fractal Geometry in Computer Science, Orsucci F., Sala N. (eds.), *Reflexing Interfaces: The Complex Coevolution of Information Technology Ecosystems,* Information Science Reference, IGI Global, Hershey, New York, 2008, pp. 308 – 328.

Saleri R. (2002) Pseudo-urban automatic pattern generation, *Proceedings 5th Generative Art Conference,* Milano, Italy, available: http://www. generativeart.com/on/cic/papersGA2002/24.pdf

Saleri, R. (2006a) Urban and architectural 3D fast processing *Proceedings 9th Generative Art Conference,* Milano, Italy, available: http://www. generativeart.com/on/cic/ papersGA2006/ 40.htm

Saleri, R. (2006b). Pseudo-Urban Automatic Pattern Generation, Sala, N. (ed.), *Chaos and Complexity in Arts and Architecture*, Nova Science, New York, pp. 161-169.

Saleri, R. (2008), A 3D Automatic Processing Of Architectural and Urban Artifacts. Orsucci, F.and Sala, N. (eds.) *Reflexing Interfaces: The Complex Coevolution of Information Technology Ecosystems*, IGI Group, Hershey, PA, pp. 278-289.

Salingaros, N. (1999). Architecture, Patterns, and Mathematics, *Nexus Network Journal 1*, pp. 75–85.

Salingaros, N. (2000).The Structure of Pattern Languages, *Architectural Research Quarterly 4*, pp. 149–161.

Salingaros, N. (2001) *I frattali nella nuova architettura* [Fractals in the new architecture], available: http://www. archimagazine.com/afrattai.htm

Semboloni, F. (2006). Self-Organized Cruiticality in Urban Spatial Development, Sala, N. (ed.), *Chaos and Complexity in Arts and Architecture*, Nova Science, New York, pp. 135-145.

Spiller, N. (2004) *A brief introduction to Marcos Novak*, Retrieved 20 June, 2009 from: http://sls2000.lcc.gatech.edu/bioblurb2.htm

Thomas, I., Tannier, C., and Frankhauser, P., (2008). Is there a link between fractal dimension and residential environment at a regional level? *Cybergeo: European Journal of Geography* [online], *Systèmes, Modélisation, Géostatistiques* [Systems, Modeling, Geostatistics], document 413, mis en ligne le 25 février 2008. URL: http://cybergeo.revues.org/index16283.html.

Witten, T.A., and Sanders, L.M. (1981). Diffusion-Limited Aggregation, a Kinetic Critical Phenomenon, *Phys. Rev. Lett.* 47, pp. 1400 – 1403.

Whyte, J. (2002). *Virtual Reality and the Built Environment*, Architectural Press, Oxford.

Wright, D.J. (1996). *Designing IFS's: the Collage Theorem*, retrieved, September 30, 2005, from: http://www.math.okstate.edu/mathdept /dynamics/lecnotes/node47.html

In: Chaos and Complexity in the Arts …   ISBN: 978-1-53612-995-3
Editors: N. Sala and G. Cappellato   © 2018 Nova Science Publishers, Inc.

Chapter 6

COMPLEXITY AND ARCHITECTURE

Nicoletta Sala[*]

Accademia di Architettura, Università della Svizzera italiana,
Mendrisio, Switzerland

ABSTRACT

The complexity is the property of a real world system that is manifest in the inability of any one formalism being adequate to capture all its properties. The complexity is also the theory of how emergent organization may be achieved by the interaction with components pushed far from equilibrium (by increasing matter, information, or energy) to the threshold between order and disorder (chaos). This threshold is where the system often interacts in a new non-linear way.

Can the complexity involve the architecture? An answer could be that architecture finds inspiration observing the nature, and nature is fractal and complex. Modern architects study the complexity and the fractal geometry to create a new kind of buildings or to understand the problems connected to the networks' organizations and to the urban growths. The aim of this paper is to present an approach that studies the complexity applied in architecture.

Keywords: box-counting dimension, complexity, fractal geometry, non-linear architecture, networks, self-similarity

---

[*] Corresponding Author Email: nicoletta.sala@usi.ch.

INTRODUCTION

The complexity is the most difficult area of chaos, and it describes the complex motion and the dynamics of sensitive systems. Sporns (2007) defines the complexity as "... the degree to which components engage in organized structured interactions."

Huberman and Hogg (1986) propose the complexity as a mixture of order and disorder (Figure 1).

The complexity is connected to the chaos which reveals a hidden fractal order underlying all seemingly chaotic events. It can occur in natural and man-made systems, as well as in meteorological systems, human beings and social structures. Complex dynamical systems may be very small or very large, and in some complex systems small and large components exist in co-operative way.

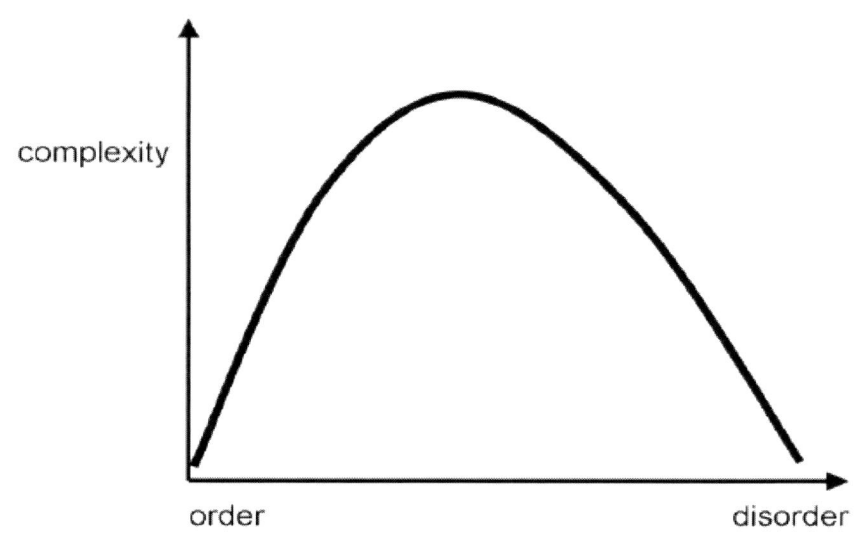

Figure 1. Complexity as a measure of order and disorder

George Birkhoff (1884-1944), one of the greatest American mathematician of the early 20th century, proposed a measure of beauty connected to the complexity. In his *Aesthetic Measure* (1933) he described a mathematical theory of aesthetics based on a simple relationship:

$$\mathbf{M} = \frac{\mathbf{O}}{\mathbf{C}}$$

(1)

where by M stands for "aesthetic measure" (or beauty), O for order and C for complexity. This measure suggests the idea that beauty has something to do with order and complexity.

Birkhoff's theory is interesting, because it is an attempt to apply complexity in the aesthetic measures. In the past, mathematical and geometrical components (for example, the golden ratio, the symmetry, the Fibonacci's sequence, and the Euclidean geometry) were present in arts and in architecture for searching an aesthetic sense and the perfection (Ghyka, 1977; Blackwell, 1984; Mamman, 1988; Hargittai, I. and Hargittai, 1996; Dunlap, 1998; Sala and Cappellato, 2003).

We can also observe the architecture using a different point of view, for example we can find some fractal or complex components that are present in the buildings or in the projects' organizations (Venturi, 1992; Eaton, 1998; Jencks, 1998; Sala, 2002; Sala and Cappellato, 2004; Sala, 2007).

Robert Venturi affirms, in his book entitled *Complexity and contradiction in architecture* (1992): "The recognition of complexity in architecture does not negate what Louis Kahn has called 'the desire for simplicity'. But aesthetic simplicity which is a satisfaction to the mind derives, when valid and profound, from inner complexity" (Venturi, 1992, p. 17).

The complexity and the fractal geometry appear in architecture in different ways, for example for reproducing the natural patterns and the natural shapes (Jencks, 1998; Sala and Cappellato, 2004).

Our complex analysis in architecture has been divided in two parts:

- on a small scale analysis (e. g., to determine the complex components present in the building shape);
- on a large scale analysis (e.g., to study the urban growth and the urban development, or to analyse the organisation of the landscape).

In this paper we will present a small scale analysis which we have organised in three parts:

1. the complexity in the buildings shape (e.g., the complex surfaces and the complex textures in the works of Franck O. Gehry and in the projects realised by Paolo Portoghesi);
2. the complexity in the building's organization (for example in analogy to the networks' organization);

3. the box-counting dimension of a design, to determine its degree of complexity (for example applied to the Mesoamerican architecture and to the modern architecture).

THE COMPLEXITY IN THE BUILDINGS SHAPE

A new kind of architecture was born by the complexity: non-linear architecture (Jencks, 1998). To research the complexity in the buildings, we can observe the complex textures (e.g., in Gehry's works) and the complex surfaces (e.g., in the Gehry's or Portoghesi's projects) derived by the observation of the nature, and the natural phenomena.

Gehry is one of the most inventive and pioneering architects working today. He has used the complexity and the fractal geometry in his recent works. Since his fish lamps of 1983, Gehry has applied the property of the self-similarity to realise the complex textures of his buildings. The self-similarity is a fractal geometry's property that permits to an object to repeat its shape in different scales. This is evident observing the metal shingle, repeated in different scales, that covers the Vitra Headquartes (1989-1992), and the Guggenheim Museum, Bilbao (1992-1997). The Figures 2 shows the skin of a snake, where it is clear the complexity and the self-similarity. The Figure 3 illustrates a portion of the Guggenheim Museum's skin. The analogy with the nature is amazing.

Figure 2. The skin of the snakes is self-similar.

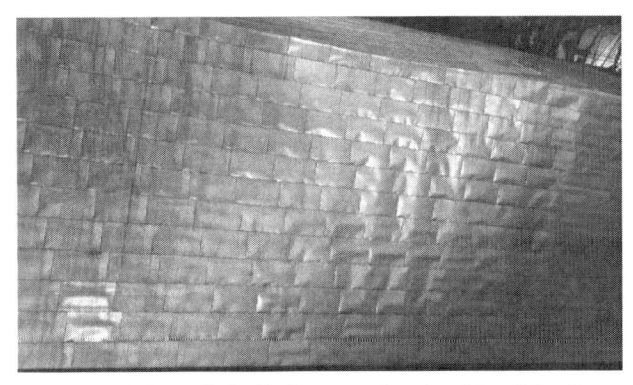

Figure 3. Gehry has used the self-similarity to realise the skin of the Guggenheim Museum, Bilbao (1992-1997).

Austrian architect Gunther Domenig realised some buildings which recall complex and fractal shapes; an example is Steinhaus (1998-2008) at Lake Ossiach (Carinthia, Austria), in Figure 4. This house has an opened outward planimetry in which are inserted distorted volumes of cement or steel, starting from a glazed cylinder which directly emerges from the subsoil.

In his project Domenig tries to capture a moment of initial chaos, therefore his work is not ended and in fact at a first glance appears "destroyed". The structure grew year by year, piece by piece, following an every-evolving set of sketches and technical drawings.

Figure 4. Gunther Domenig: Steinhaus (1998-2008) at Lake Ossiach (Carinthia, Austria).

Italian architect Paolo Portoghesi, inspired by the chaotic movements connected to the gas or liquid motion, has realised the Hotel Savoia (1992-1996) in Rimini (Italy) that contains smoothed surfaces inspired by the wave motion as a metaphor of the sea (shown in Figure 5). The hotel is situated on the Adriatic (Portoghesi, 1999). Figure 6 illustrates a particular of the model of Hotel Savoia. Portoghesi used the fractal geometry, in particular the self-similarity, in the Casa Baldi, Rome (1959), in the Villa Papanice, Rome (1966), and in the Islamic Cultural Center and Mosque in Rome (1975) (Portoghesi, 1999). His more notable works include Casa Andreis (Sandriglia, Rieti, 1964), the Istituto Tecnico Industriale (Aquila, 1969), the Chiesa della Sacra Famiglia (Fratte, Salerno, 1969), the City Library and Social Center (Avezzano, 1970), the Royal Palace of Amman, Jordan (1973) and the Urban Planning Scheme and International Airport for Khartoum, Sudan (1973).

Gehry and Portoghesi have developed a non-linear architecture in conscious way.

Figure 5. The wave motion is an example of complex system.

Figure 6. Portoghesi's Hotel Savoia (Rimini, Italy).

THE COMPLEXITY IN THE BUILDINGS ORGANIZATION

Architecture finds inspiration by the observation of biological systems. Salingaros (2004) affirms: "The idea of a biological connection has been used in turn by traditional architects, modernists, postmodernists, deconstructivists, and naturally, the 'organic form' architects. One might say that architecture's proposed link to biology is used to support any architectural style whatsoever. When it is applied so generally, then the biological connection loses its value, or at least becomes so confused as to be meaningless."

The complexity is also connected to the network theory (Giuliani, 2008; 2012). It does not describe the quality of the topology of certain network (such as number and kind of network connections) but also the relations neighbouring systems and the embedding into the total organism. In particular, Sporns et al (2000) affirm: "…the graph structure of the interactions, in a network topology, places important constraints on the system's dynamics, by directing information flow, creating patterns of coherence between components, and by shaping the emergence of macroscopic system states. Complexity is sensitive to changes in network topology" (Sporns et al., 2000). Wolfgang Höhl (2004) tries to identify useful organisational patterns in biological and technological networks as an approach for future network planning. He compares biological and networks due to the system behaviour, organization, environment, structure, function and form. Höhl divides the buildings in two types, in according to their construction, obtaining: massive and skeleton constructions.

In particular, he observes that: "in skeleton construction we will find so-called 'organic backbones networks': homogeneous and harmonically networks with a central "spinal cord". In massive constructions we can observe so-called: "complex exuberant networks", a heterogeneous growing network with un-used rudiments" (Höhl, 2004, p. 60).

THE BOX-COUNTING DIMENSION

The box-counting dimension is connected to the problem of determining the fractal dimension of a complex two-dimensional image. It is defined as the exponent D_b in the relationship:

$$N(d) \approx \frac{1}{d^{D_b}} \tag{2}$$

where $N(d)$ is the number of boxes of linear size d, necessary to cover a data set of points distributed in a two-dimensional plane. The basis of this method is that, for objects that are Euclidean, equation (2) defines their dimension.

One needs a number of boxes proportional to $1/d$ to cover a set of points lying on a smooth line, proportional to $1/d^2$ to cover a set of points evenly distributed on a plane, and so on. Applying the logarithms to the equation (2) we obtain: $N(d) \approx -D_b \log(d)$. The box-counting dimension can be produced using this iterative procedure:

- superimpose a grid of square boxes over the image (the grid size as given as s_1);
- count the number of boxes that contain some of the image ($N(s_1)$);
- repeat this procedure, changing (s_1), to smaller grid size (s_2);
- count the resulting number of boxes that contain the image ($N(s_2)$);
- repeat this procedures changing s to smaller and smaller grid sizes.

The box-counting dimension is defined by:

$$D_b = \frac{[\log(N(s_2)) - \log(N(s_1))]}{\left[\log\left(N\left(\frac{1}{s_2}\right)\right) - \log\left(N\left(\frac{1}{s_1}\right)\right)\right]} \tag{3}$$

where $1/s$ is the number of boxes across the bottom of the grid.

We can apply the box-counting dimension in architecture, too. It is calculated by counting the number of boxes that contain lines from the drawing inside them. Next Figure 7 illustrates the box count for the elevation of a Frank Llyod Wright's building (Robie House, 1909) (Bovil, 1996, p. 122). Table 1 contains the number of boxes counted, the number of boxes across the bottom of the grid, and the grid size. The box-counting dimension of Robie House, calculated using (3), is a value between 1.441 and 1.485.

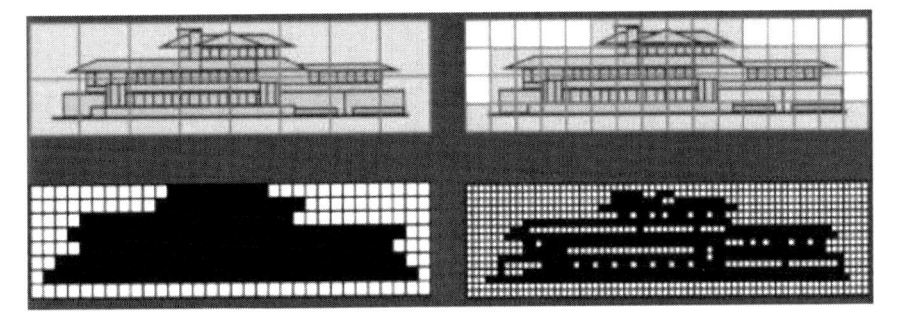

Figure 7. Box-counting method applied to Wright's Robie House.

Table 1. Number of boxes counted, the number of boxes across the bottom of the grid, and the grid size

Box count	Grid size	Grid dimension
16	8	24 feet
50	16	12 feet
140	32	6 feet
380	64	3 feet

To determine the degree of the complexity in the Mesoamerican arts and architecture, Burkle-Elizondo et al. have collected more than a hundred of images of Mesoamerican artistic and architectural works by reviewing literature on archeology (Burkle-Elizondo and Cepeda-Valdéz, 2002; Burkle-Elizondo et al., 2003; 2004).

All these images have been digitized using a Printer-Copier-Scanner (Hewlett Packard®, Model LaserJet 1100A) and saved in bitmap (*.bmp) format on a Personal Computer (Hewlett Packard®, Model Pavilion 6651). Thereafter, these images were analyzed with the program Benoit®, version 1.3 in order to calculate Box (D_b), Information (D_i), and Mass dimensions (D_M), and their respective standard errors and intercepts on log-log plots. It was taken under consideration that the information dimension differs from the box dimension, because its boxes contains more points. For all the cases the fractal dimension values were high from a $D_b = 1.803\pm0.023$ for the left and superior side of the "Vase of seven gods", to a $D_M = 2.492\pm0.195$ for the left side of the "Door to underworld of the Temple 11, platform" at Copán. The degree of complexity found in the Mesoamerican buildings can be explained if we remember the basis of Mesoamerican Cosmo vision, and the idea how the Universe works, that have influenced this architecture (Burkle-Elizondo et al., 2004).

CONCLUSION

The complexity paradigm has developed two different traditions: one in architecture, another in science. In this work we have described only an approach where the complexity has been analysed using three different points of view:

1. to search the complexity in the buildings shape
2. to find the complexity in the building's organization
3. to determine the degree of complexity in a building
    (using the box-counting dimension).

The examples described in the parts 1 and 2 of this paper introduce the concept of non-linear architecture. Non-linear architecture has influenced Peter Eisenman's Aroff Center in Cincinnati, Frank Gehry's Guggenheim Museum in Bilbao and Daniel Libeskind's Jewish Extension to the Berlin Museum. All three buildings were partly generated by non-linear methods.

The complexity in architecture is also connected to the networks.

As Robert Venturi (1992) connects the complexity to the concept of contradiction: "I like complexity and contradiction in architecture. I do not like the incoherence or arbitrariness of incompetent architecture not the precious intricacies of picturesqueness or impressionism. Instead, I speak of a complex and contradictory architecture based on the richness and ambiguity of modern experience, including that experience which is inherent in art. Everywhere, except in architecture, complexity and contradiction have been acknowledged, from Gödel's proof of ultimate inconsistency in mathematics to T.S. Eliot's analysis of "difficult" poetry and Joseph Albers' definition of the paradoxical quality of painting" (Venturi, 1992, p. 16).

Will the complexity paradigm influence the architecture?

Quoting Alexander (2002-2005): 'People used to say that just as the twentieth century had been the century of physics, the twenty-first century would be the century of biology... We would gradually move into a world whose prevailing paradigm was one of complexity, and whose techniques sought the co-adapted harmony of hundreds or thousands of variables. This would, inevitably, involve new technique, new vision, new models of thought, and new models of action. I believe that such a transformation is starting to occur.... To be well, we must set our sights on such a future.'

REFERENCES

Alexander , C. (2002-2005). *The Nature of Order* (4 voll.), CES, Berkeley.

Birkhoff, G. D. (1933). *Aesthetic Measure*. Harvard University Press, Cambridge, MA.

Blackwell, W. (1984). *Geometry in Architecture*, John Wiley and Sons, London.

Bovill, C. (1996). *Fractal Geometry in Architecture and Design*, Birkhäuser, Boston.

Burkle-Elizondo, G. and Cepeda-Valdéz R. (2002). Do The Artistic and Architectural Works Have Fractal Dimension? *Emergent Nature*, M.M. Novak (ed.), World Scientific, Singapore, pp. 431-432.

Burkle-Elizondo, R., Sala, N. and Cepeda-Valdéz, R. (2003). Complexity in the Mesoamerican artistic and Architectural Works, *Programme and Abstracts International Nonlinear Sciences Conference (Research and Applications in the Life Sciences)*, Vienna, Austria, p. 18.

Burkle-Elizondo, R., Sala, N. and Cepeda-Valdéz, R. (2004). Geometric And Complex Analyses Of Maya Architecture: Some Examples. *Nexus V: Architecture and Mathematics*, ed. K. Williams K., Edizioni Dell'Erba, Fucecchio, Italy, pp. 57-68.

Dunlap R.A., *The Golden Ratio and Fibonacci Numbers*, World Scientific, Singapore, 1998.

Eaton, L.K. (1998). Fractal Geometry in the Late Work of Frank Llyod Wright: the Palmer House. *Nexus II: Architecture and Mathematics*, ed. K. Williams, Edizioni Dell'Erba, Fucecchio, Italy, pp. 23 - 38.

Ghyka, M. (1977). *The Geometry of Art and Life*, Dover Publications, USA.

Giuliani, A. (2008). Networks: Uses and Misuses of an Emergent Paradigm. Orsucci, F., and Sala, N. (eds) *Reflexing Interfaces: The Complex Coevolution of Information Technology Ecosystems*, IGI Global, Hersey, pp. 174 – 184.

Giuliani, A. (2012). Networks: A Sketchy Portrait Of An Emergent Paradigm. Orsucci, F., and Sala, N. (eds) *Complexity Science, Living Systems and Reflexing Interfaces: New Models and Perspectives*, IGI Global, Hersey, (in print)

Hargittai, I. and Hargittai, M. (1996). *Symmetry: A Unifying Concept*, Random House, New York.

Höhl, W. (2004). Neural networks an information interchange in buildings, Collins, M.W. and Brebbia, C.A. (eds) *Design and Nature III*, WIT Press,

Southampton, Boston, pp. 55-62 http://www.katarxis3.com/Salingaros-Biological_Understanding.htm.

Huberman, B.A., Hogg, T. (1986). Complexity and adaptation. *Physica D* 22, 376-384.

Jencks, C. (1998). Complexity definition and nature's complexity, *Architectural Design*, n. 129, pp. 8-10.

Portoghesi, P. (1999). *Natura e Architettura*, Skira, Milano (English version: *Nature and Architecture*, Skira, Milano, 2000).

Sala, N. (ed.) (2007). *Chaos and Complexity in Arts and Architecture*, Nova Science, New York.

Sala, N. and Cappellato, G. (2003). *Viaggio matematico nell'arte e nell'architettura* [Mathematical journey in art and architecture], Franco Angeli, Milano.

Sala, N. and Cappellato, G. (2004). *Architetture della complessità* [Architectures of complexity], Franco Angeli, Milano.

Sala, N., (2002). The presence of the Self- Similarity in Architecture: Some examples. *Emergent Nature*, M.M. Novak (ed.), World Scientific, Singapore, pp. 273 – 283.

Salingaros, N. (2004). Towards A Biological Understanding of Architecture and Urbanism: Lessons From Steven Pinker. *Katarxis N° 3*. Retrieved, 10 March 2011, from: http://www.katarxis3.com/Salingaros-Biological_Understanding.htm

Sporns, O. (2007), *Scholarpedia*, 2(10):1623.

Sporns, O., Tononi, G., and Edelman, G.M. (2000). Theoretical neuroanatomy: Relating anatomical and functional connectivity in graphs and cortical connection matrices. *Cereb. Cortex* 10, 127-141.

Venturi, R. (1992). *Complexity and contradiction in architecture*, The Museum of Modern Art, New York.

In: Chaos and Complexity in the Arts …   ISBN: 978-1-53612-995-3
Editors: N. Sala and G. Cappellato   © 2018 Nova Science Publishers, Inc.

Chapter 7

VEDIC FRACTALS

Mamta Rani[1], Sanjaya Tripathi[2] and Arun Prakash Agarwal[3]*

[1]Krishna Engineering College, Ghaziabad, India
[2]Galgotias College of Engineering and Technology, Greater Noida, India
[3]Amity University, Noida, India

ABSTRACT

Professor Mandelbrot founded fractal geometry in 1975 and included many old classical mathematical models in the gallery of fractal in 1970s, e.g., Pascal triangle (1654), Cantor set (1883), Koch curve (1904), Sierpinski Fractals (1916), Julia sets (1918) etc. Our study shows that germs of fractals exist in old Indian literature, e.g., fractal architecture in Indian temples and fractal weapons. The purpose of this paper is to collect a few examples from old Indian history and present their fractal aspects.

Keywords: fractals, vedic fractals

---

* Corresponding Author Email: mamtarsingh@gmail.com.

1. Introduction

Benoit B. Mandelbrot, a French and American mathematician, is the best appreciated for his first broad attempt to describe irregular shapes in nature. He founded fractal geometry in 1975, which impacts mathematics [2, 5, 7], diverse sciences [1, 8, 15, 19], and arts [6, 9, 20].

Nature is filled with complex geometrical shapes such as biological shapes, branching patterns of rivers, seashore lines, and branching pattern of trees. All these complex shapes have self-similarity feature. Professor Mandelbrot discovered that self-similarity is the universal property that underlies such complex shapes, and he coined the expression "fractal". Furthermore, he has illustrated its properties mathematically and founded a new methodology for analyzing complex systems [12]. Also, see [11].

Professor Mandelbrot included many old classical mathematical models in the gallery of fractal in 1970s, e.g., Pascal triangle (1654), Cantor set (1883), Koch curve (1904), Sierpinski Fractals (1916), Julia sets (1918) etc. [14]. Our study shows that germs of fractals exist in old Indian literature, e.g., fractal architecture in Indian temples and fractal weapons. The very old period of Indian history is called as vedic period, so the fractals in vedic period will be called as vedic fractals. The purpose of this paper is to collect such type of examples in vedic period and look upon them as fractal.

2. Vedic Fractals

"As from a blazing fire thousands of sparks fly forth, each one looking self-similar to its source, so from the Eternal comes a great variety of things, and they all return to the Eternal finally."

… Mundaka Upanishad II.1.1 (cf. [10])

Indian heritage is enriched by fractals. Many objects in old Indian literature have fractal shapes and can be generated by mathematical formulations using recursive algorithms. These objects may be analyzed again from the fractal aspect. Here, we give a collection of such type of objects, may be called as vedic fractals, and their generation methods.

2.1. Meru

Pingalacharya came out with an arrangement of numbers in such a way that one may get various coefficients in the binomial expansion of $(x + 1)^n$. He named this arrangement of numbers as Meru (piece of mountain).

See Figure 1 showing the arithmetic triangle for the binomial coefficients for $n = 7$. Notice that, n^{th} row has $n + 1$ entries. For example, with $n = 4$,

$$(x+1)^4 = x^4 + 4x^3 + 6x^2 + 4x + 1.$$

The row number 4 in Figure 1 reads 1, 4, 6, 4, 1.

```
1
1 1
1 2 1
1 3 3 1
1 4 6 4 1
1 5 10 10 5 1
1 6 15 20 15 6 1
1 7 21 35 35 21 7 1
```

Figure 1. Bionomial Expansion in the form of mountain for $n = 7$.

In 1150 AD, Bhāskarācārya discussed Meru in his book Lilāvati, which have been used as a textbook in India for many centuries [13]. Chu Shih Chieh, a Chinese mathematician, also came up with the same bionomial arrangement in 13th century [3]. Meru was reintroduced by a French mathematician Blaise Pascal in 1654 [14]. Meru is now, generally, known as Pascal triangle. Pascal gave color-coding in the Meru by surrounding each coefficient by a small hexagon and color each hexagon of even coefficient as white and of odd coefficient as blue/black.

See Meru in Figure 2 when n = 15.

Figure 2. Meru.

2.2. Fractal Weapons

Hindu's mythological divine weapon Sudarśana Chakra and Trident (Trishool) is a fractal object.

Sudarśana Chakra
Sudarśana Chakra is produced as a fractal projective hedgehog due to a function, which is continuous everywhere but differential nowhere. Following is the equation due to Rani & Kumar [17], responsible for construction of Sudarśana Chakra.

Theorem. There exists a fractal hedgehog $H_h \subset R^2$ if $q(\theta) = h\ (\cos\theta, \sin\theta)$ is a function of the form

$$q(\theta) = \sum_{1}^{\infty} (1/\alpha^n) \sin\ (\beta^n \theta)$$

$$\tag{1}$$

where β is an odd integer and α a positive real number such that $\alpha > |\beta|$ and $\beta^2 > \alpha(1 + 3\pi/2)$, then the hedgehog satisfies the following two properties:
1. the curve H_h is continuous but nowhere differentiable;
2. the curve H_h has infinite length.

Sudarsana Chakra is obtained by putting $n = 2$, $\alpha = 2005$ and $\beta = -2001$ in Eq. (1). See the fractal Sudarsana Chakra in Figure 3.

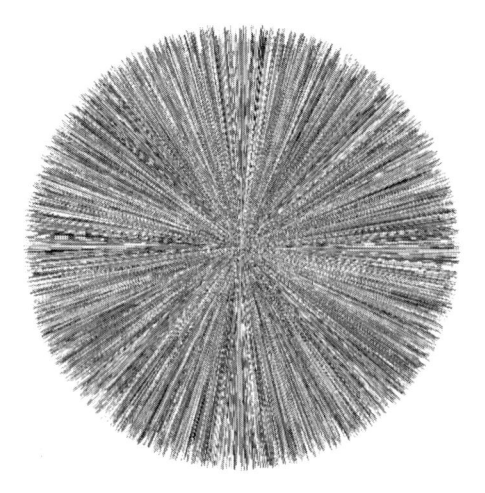

Figure 3. Sudarsana Chakra: A fractal projective hedgehog.

Trident (Trishool)

Another Hindu's mythological divine weapon Trident (Trishool) is a Julia set produced by the Carotid Kundalini function, abbreviated as C-K function. The mathematical formulation of the C-K complexed function due to Cooper [4] is $K_{N,c}(z) = \cos(N\, z\, cos^{-1}(z)) + c$, where N and c are parameters and, as usual, $\cos z = \frac{(e^{iz}+e^{-iz})}{2}$

Following is the iterated Carotid-Kundalini (C-K) equation due to Rani and Negi [18].

$$z_{j+1} = s\left(K_{N,c}(z_j)\right) + (1-s)z_j, \text{ where } 0 \le s \le 1 \qquad (2)$$

Filled superior C-K Julia set obtained from Eq. 2 for $N = 0.85$, $s = 0.01$, $c = -9.7$ is akin to the Trident (Figure 4).

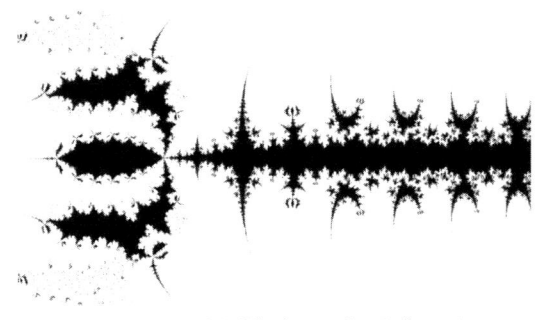

Figure 4. Trident (A Carotid Kundalini filled superior Julia set).

2.3. Fractal Architecture in Temples

Indian and Southeast Asian temples and monuments exhibit a fractal structure: a tower surrounded by smaller towers, surrounded by still smaller towers, and so on, for eight or more levels. William Jackson and other researchers noticed fractal architecture in Indian temples [9, 10]. Kanadariya temple in Khajuraho, Ramanatha Swamy temple in Rameshwaram, Trymbakeshwar temple in Nashik, Meenakshi temple in Madurai are just few examples of fractal architecture (see Fig. 5). To see more examples of fractal temple, one may refer to [9, 10] and several cross references thereof.

Fractal architecture of Kanadariya Temple in Khajuraho, North India

A temple in Banaras, North India with self-similar spires composed of smaller and smaller spires

Ramanatha Swamy temple in Rameshwaram, South India

Meenakshi Temple in Madurai, Souh IndiaTrymbakeshwar temple, Nashik, India

Figure 5. Fractal architecture in temples.

2.4. Lord Ganesha

Julia sets can be generated for the complex polynomial $A_c = \bar{z}^d + C$, where $d \geq 2$, with respect to superior iterations. Superior iterations are formulated as $z_n = \beta_n f(z_{n-1}) + (1-\beta_n) z_{n-1}$, where z is a complex number, $0 < \beta_n \leq 1$ and $\{\beta_n\}$ is convergent to a non-zero number. Escape criterion for $A_c = \bar{z}^d + C$ with respect to superior iterates is max $\{|c|, (2/\beta)^{1/(d-1)}\}$, where $0 < \beta \leq 1$.

At $d = 3$, Julia sets for $(\beta, c) = (0.5, 1.0+2.13i)$ and $(\beta, c) = (0.5, -0.9+2.2i)$ closely resemble to the Hindu's mythological elephant headed Lord Ganesha (Figure 6) [16].

Figure 6. Two views of elephant headed Lord Ganesha.

CONCLUSION

The study in this paper shows that the mathematical models recognized by Mandelbrot as classical examples of fractals are not the oldest examples of fractals. There are many fractal shapes that exist in very old Indian time period (vedic period). Such fractal may be called as vedic fractals.

REFERENCES

[1]   Michael F. Barnsley, and Hawley *Rising, Fractals Everywhere* (2nd ed.), Academic Press Professional, Boston, 1993.

[2]   John Briggs, Fractals: *The Patterns of Chaos* (2nd ed.), Thames and Hudson, London, 1992.

[3]   J. L. Coolidge, *The story of the bionomial theorem*, 1949, 147-157. http://poncelet.math.nthu.edu.tw/disk5/js/geometry/binomial.pdf

[4]   G. R. J. Cooper, Julia sets of the complex Carotid-Kundalini function. *Computer Graphics*, 25, 2001,153-158.

[5]   Robert L. Devaney, A First *Course in Chaotic Dynamical Systems: Theory and Experiment,* Westview Press, CO, 1992.

[6]   Ron Eglash, A*frican Fractals: Modern Computing and Indigenous Design,* Rutgers University Press, New Brunswick, 1999.

[7]   Kenneth Falconer, *Fractal Geometry: Mathematical Foundations and Applications* (3rd ed.), John Wiley & Sons Ltd., NY, 2003.

[8]   R. Hohlfeld, and N. Cohen, Self-similarity and the geometric requirements for frequency independence in antenna, *Fractals*, 7(1), 1999, 79-84.

[9]   William J. Jackson, *Hindu temple fractals.*

[10]   http://liberalarts.iupui.edu/~wijackso/tempfrac/

[11]   William J. Jackson, *Fractals in Indian architecture.*

[12]   http://classes.yale.edu/fractals/Panorama/Architecture/IndianArch/Indian Arch.html

[13]   B. Mandelbrot, *Fractals: Form, Chance and Dimension*, W.H. Freeman and Co., New York, 1977.

[14]   B. B. Mandelbrot, *The Fractal Geometry of Nature*, W.H. Freeman and Co., New York, 1982.

[15]   K. S. Patwardhan, S. A. Naimpally and S. L. Singh, Lîlâvatî of Bhâskarâcârya: A *Treatise of Mathematics of Vedic Tradition*, Motilal Banarsidass Publishers, Delhi, 2001.

[16]   H. Peitgen, H. Jürgens and D. Saupe, *Chaos and Fractals*, Springer-Verlag, New York, Inc, 1994.

[17]   Gongwen Peng and Tian Decheng, The fractal nature of a fracture surface, *J. Phys. A: Math. Gen.*, 23(14), 1990, 233-257.

[18]   M. Rani, Superior Antifractals, in: *IEEE Proc. ICCAE 2010*, Feb 26-28, vol. 1, 798-802.

[19] M. Rani and V. Kumar, A fractal hedgehog theorem, *J. Korea Soc. Math. Edu. Series B; Pure & Applied Math.*, 9(2), 2002, 91-105.

[20] M. Rani, and A. Negi, New Julia sets for complex Carotid-Kundalini function, *Chaos, Solitons, Fractals*, 36(2), 2008, 226-236. MR2382153 Zbl 1142.37347.

[21] K. M. Roskin, and J. B. Casper, *From Chaos to Cryptography*, 1998. http://xcrypt.theory.org/

[22] Richard Taylor, Adam P. Micolich, and David Jonas, *Fractal Expressionism: Can Science Be Used To Further Our Understanding Of Art?*

[23] http://phys.unsw.edu.au/phys_about/PHYSICS!/FRACTAL_EXPRESSI ONISM/fractal_taylor.html

[24] http://www.google.co.in/search?q=images+of+indian+temples&hl=en& prmd=imvns&tbm=isch&tbo=u&source=univ&sa=X&ei=A6G4T_PoNo -0rAfA-MjUDA&sqi=2&ved=0CHsQsAQ&biw=1118&bih=615

In: Chaos and Complexity in the Arts …   ISBN: 978-1-53612-995-3
Editors: N. Sala and G. Cappellato   © 2018 Nova Science Publishers, Inc.

Chapter 8

FRACTAL GEOMETRY IN VIRTUAL WORLDS AND IN SECOND LIFE

*Nicoletta Sala**

Accademia di Architettura, Università della Svizzera italiana,
Largo Bernasconi, Mendrisio, Switzerland
Dipartimento di Informatica e Comunicazione
Università dell'Insubria, Via Mazzini 5. Varese, Italy

ABSTRACT

Fractal geometry is an excellent tool for modelling natural shapes
(e.g., textures, ferns, trees, flowers, seashells, rivers, mountains), and its
important applications appear in computer science. In particular, this
"quite young" geometry permits to reproduce, in computer graphics and
in the virtual reality, the complex and irregular forms present in nature
using simple iterative or recursive instructions. The aim of this paper is to
present some recent applications for generating virtual landscapes,
territories and complex shapes in virtual worlds and in Internet based
virtual worlds, for example Second Life.

Keywords: fractal geometry, self-similarity, iterated function systems, L-
systems, computer graphics, virtual reality, virtual worlds, second life

* Corresponding Author Email: nsala@arch.unisi.ch.

1. INTRODUCTION

In the beginning of the 20th century French mathematicians Pierre Fatou (1878-1929) and Gaston Julia (1893-1978) worked on first fractal sets, but only in the middle of the last century the Polish-born Franco-American Benoit Mandelbrot coined the word "Fractal" to denote irregular shapes (Mandelbrot, 1975). During these last decades, many scholars were helped by the electronics evolution and the increase in the computer calculation power for applying fractal geometry to different disciplines (for example, from the biology to the computer science), and the multiplicity of applications had an important role in the diffusion of the fractals (Mandelbrot, 1982; Leland et al., 1993; Nonnenmacker et al., 1994; Eglash, 1999; Barnsley et al., 2002; Sala, 2006; Vyzantiadou et al., 2007; Sala, 2008).

Fractal geometry is able to describe the Nature, replacing Euclidean geometry (which dominated our mathematical thinking for hundreds of years). This paper presents some applications of fractal geometry for generating virtual landscapes, territories and complex shapes in virtual worlds and in three-dimensional Internet based virtual worlds. The paper is organized as follows: section 2 describes the concept of fractal objects. Section 3 introduces some applications of the fractal geometry for modelling 2D and 3D objects in virtual reality environments and in virtual worlds. Section 4 introduces some applications of fractal geometry in a Internet based virtual world: Second Life. Section 5 is dedicated to the conclusions.

2. FRACTAL OBJECTS: SOME EXAMPLES

Mandelbrot, the "father" of fractal geometry, described a fractal object as a fragmented geometric shape that can be subdivided in parts, each of which is approximately a reduced-size copy of the whole. This property is called: "self-similarity" (Mandelbrot, 1982).

Fractals are generally self-similar on multiple scales. So, many fractals can be described in terms of iterations or recursions.

Some objects that are now called "fractals" were discovered and explored long before the word was coined. One of the oldest image of fractal was described by Apollonius of Perga (about 300 BC) and it was his Apollonian gasket (Figure 1a). Probably, the oldest handmade fractal object is situated in the cathedral of Anagni (Italy). Inside the cathedral, built in the year 1104, there is a floor (Figure 1b) which is adorned with dozens of mosaics, each in the form of a Sierpinski triangle (also known as Sierpinski gasket, or Sierpinski sieve, and shown in the Figure 1 c) (Sala, 2000).

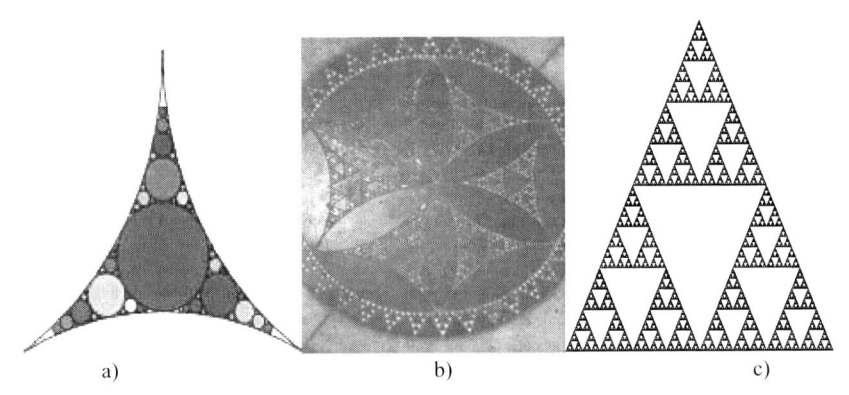

Figure 1. (a) Apollonian gasket, (b) the floor of the Cathedral of Anagni (Italy, 1104), and (c) a Sierpinski triangle.

Simple iterative rules can be used for describing three-dimensional fractal objects (for example, Sierpinski tetrahedron and the Menger cube). Figure 2 shows a Sierpinski tetrahedron (also known as Sierpinski sponge, or Tetrix,) which is the 3D version of the famous Sierpinski triangle.

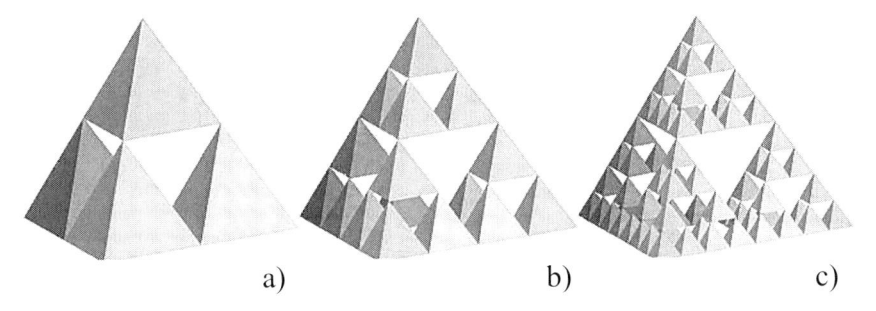

Figure 2. The first three steps of a Sierpinski tetrahedron

Figure 3 illustrates the different steps to obtain another three-dimensional fractal object: Menger cube (also known as Menger sponge), which was first described by Austrian mathematician Karl Menger (1902-1985) in 1926. The basic constructive instructions of a Menger cube are the following:

1. Take a cube (3a),
2. Divide each face of the cube into 9 squares, obtaining 27 subcubes,

3. Remove the cube at the middle of every face, and remove the cube in the center, leaving 20 subcubes (3b).
4. Repeat steps 1-3 for each of the remaining subcubes.

Figures 3c and 3d show respectively the second and the third repetition of the basic instructions. The Menger cube itself is the limit of this process after an infinite number of iterations.

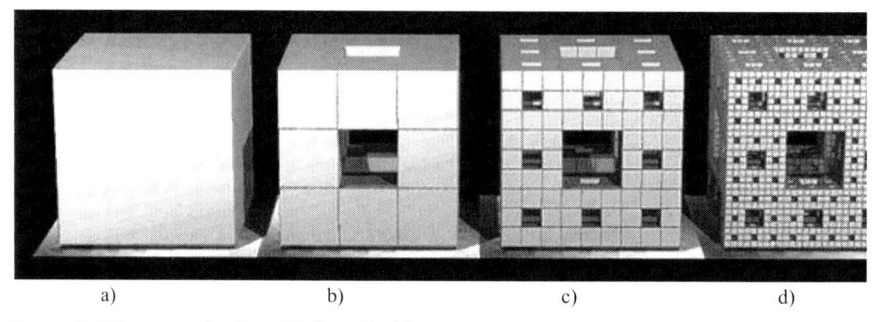

a)   b)   c)   d)

Figure 3. Menger cube is a 3D fractal object.

Complex objects in 2D and 3D which recall natural forms (e.g., trees, ferns, mountains, and rivers) are generated in computer graphics using fractal algorithms based on few iterative instructions.

3. FRACTAL GEOMETRY IN COMPUTER GRAPHICS FOR MODELING VIRTUAL WORLDS

Computer graphics is becoming a very active field of information technology science, and it applies fractal geometry, in particular the property of the self-similarity, in different ways. In this field, it is important to remark that the term "fractals" has been generalized by the computer graphics community and it includes objects outside Mandelbrot's original definition (Foley et al., 1997).

An interesting application of fractal geometry in computer science is for modeling virtual worlds including terrain, mountains, rivers, trees, flowers and other natural shapes.

Fournier et al. (1982) developed a method for generating a terrain in 3D where the basic element was a triangle. This mechanism is based on recursive subdivision for a triangle. Here, the midpoints of each side of the triangle are connected, creating four new sub-triangles, and perturbing the midpoints of the original in the direction of the normal of the triangle, as shown in Figure 4. Other polygons can be used to generate the grid (e.g., squares and hexagons), as illustrated in Figure 5. This process, when iterated few times, produces a deformed grid which represents a surface, and after the rendering phase a realistic fractal mountain appears (as shown in Figure 6).

Using this method two difficulties emerge. These are classified by Fournier as internal and external consistency problems (Fournier et al., 1982, pp. 374-375). Internal consistency requires that the shape generated should be the same irrespective of the orientation of the objects that are involved in the subdivision.

External consistency concerns the midpoint displacement at shared edges and their direction of displacement. This process, when iterated, produces a deformed grid which represents a surface.

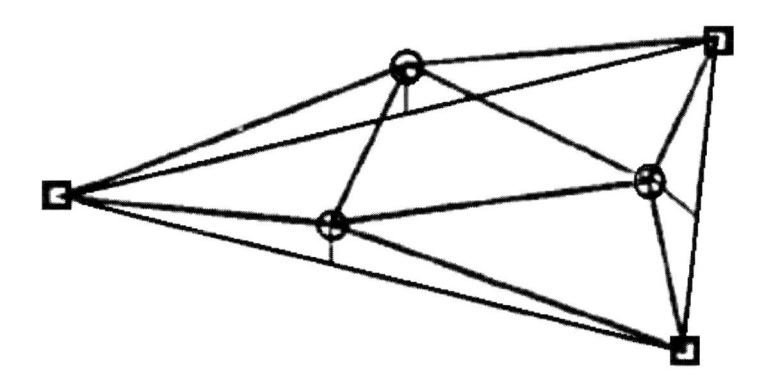

O New points □ Original points

Figure 4. Triangle subdivision.

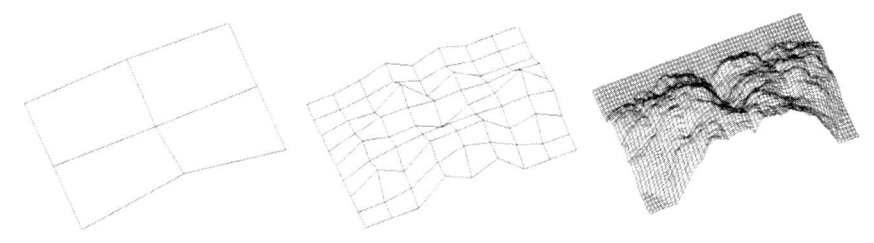

Figure 5. Grid of squares generated by a recursive subdivision and applying the fractional Brownian motion.

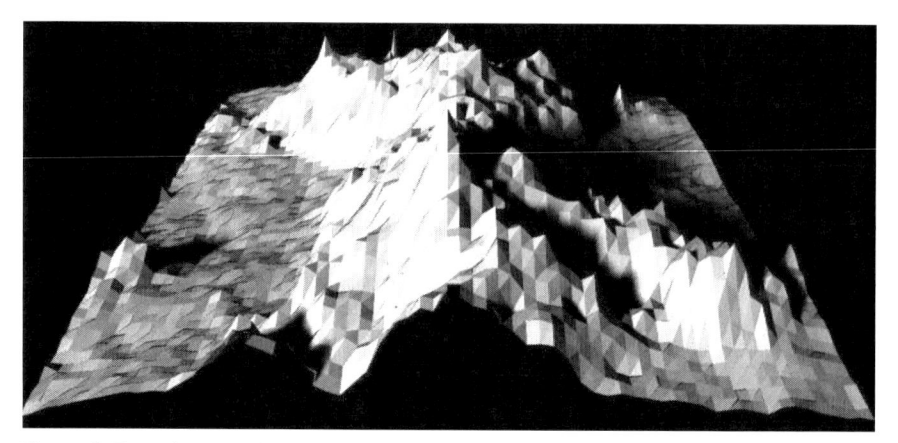

Figure 6. Fractal mountains generated using the recursive subdivisions.

Prusinkiewicz and Hammel (1993) combined the midpoint-displacement method for mountain generation with a model of a non-branching river originated by Mandelbrot (Mandelbrot, 1978; 1979).

Maràk et al. (1997) proposed a method for synthetic terrain erosion, that was based on rewriting process of matrices representing terrain parts. They found a rewriting process which was context sensitive.

Plant and tree models are fundamental for representing gardens, forests and interior scenes. They can be generated using fractal algorithms based on Iterative Function Systems.

Figure 7 shows a fern-like image (Barnsley's fern) created using the IFS which is produced by rectangles, that are put in one another. The first point is drawn at the origin ($x_0 = 0$, $y_0 = 0$) and the new points are computed in random and iterative way by applying one of the following four coordinate transformations (Barnsley, 1993). In the detail, first transformation:

$$\begin{cases} x_{n+1} = 0 \\ y_{n+1} = 0.16\ y_n. \end{cases}$$

This coordinate transformation, that draws the stem, is chosen 1% of the time and maps any point in the segment line drawn in green in the Figure 7.

Second transformation:

$$x_{n+1} = 0.2\ x_n - 0.26\ y_n$$

$$y_{n+1} = 0.23\ x_n + 0.22\ y_n + 1.6.$$

This coordinate transformation draws the bottom frond on the left, and it is chosen 7% of the time maps any point inside the black rectangle to a point inside the red rectangle in the Figure 6.

Third transformation:

$$\begin{cases} x_{n+1} = -0.15\ x_n + 0.28\ y_n \\ y_{n+1} = 0.26\ x_n + 0.24\ y_n + 0.44. \end{cases}$$

This coordinate transformation draws the bottom frond on the right. It is chosen 7% of the time and maps any point inside the black rectangle to a point inside the dark blue rectangle in the Figure 6.

Fourth transformation:

$$\begin{cases} x_{n+1} = 0.85\ x_n + 0.04\ y_n \\ y_{n+1} = -0.04\ x_n + 0.85\ y_n + 1.6. \end{cases}$$

This coordinate transformation generates successive copies of the stem and bottom fronds to make the complete fern. It is chosen 85% of the time and maps any point inside the black rectangle to a point inside the red rectangle in the Figure 7. The fern is inside the following range: $-2.1818 \le x \le 2.6556$ and $0 \le y \le 9.95851$.

Figure 7. Fern-like (Barnsley's fern) computed using the IFS. The fern has high degree of similarity to real one.

In 1968, the Hungarian biologist Aristed Lindenmayer (1925-1989) proposed the formalism of L-systems which concerned in a mathematical theory of plant development (Lindenmayer, 1968). An L-system is a parametric rewriting system operating on branching structures represented as bracketed strings of modules. In the context of L-systems, the term "module" represents any discrete constructional unit which is repeated as the plant develops, for example a branch or a flower. Thus, L-systems provided a formal description of the development of such simple multicellular organisms, and they also illustrated the neighbourhood relationships between plant cells.

L-system can be also defined as a formal grammar (a set of rules and symbols) most famously used for modelling the growth processes of plant development, and it has been thought able for modelling the morphology of a variety of organisms. The differences between L-systems and Chomsky grammars are well described by Prusinkiewicz and Lindenmayer that affirmed: "The essential difference between Chomsky grammars and L-systems lies in the method of applying productions. In Chomsky grammars productions are applied sequentially, whereas in L-systems they are applied in parallel and simultaneously replace all letters in a given word. This difference highlights the biological motivation of L-systems. Productions are intended to capture cell divisions in multicellular organisms, where many divisions may occur at the same time. Parallel production application has an essential impact on the formal properties of rewriting systems" (Prusinkiewicz and Lindenmayer, 1990, pp. 2 – 3).

Prusinkiewicz extended L-systems in a manner suitable for simulating the wide range of interactions between a developing plant and its environment, and proposed Open L-Systems (Mech and Prusinkiewicz, 1996). It is important to remark that the development can be controlled by lineage (in context-free, or OL-systems) and by endogenous interaction (in context-sensitive, or IL-systems).

Later, L-systems were extended for plant modelling, and plant modelling emerged as various area of interdisciplinary research, such as mathematical, biology, medicine, plant science, and computer science.

Figure 8 shows the geometric model of three-dimensional lilac inflorescences (8a), the infloscerence skeleton without flowers (8b) and the rendering which is like real one (8c) (Prusinkiewicz and Lindenmayer, 1990, p. 92 and p. 93) .

(a)   (b)   (c)

Figure 8. (a) A decussate branching pattern of three-dimensional lilac inflorescences model, (b) the inflorescence skeleton without flower and (c) the rendering.

In the work *Plants, Fractals, and Formal Languages* Smith (1984) proved that L-systems were useful in computer graphics for describing the structure of certain plants and trees. He introduced a new class of fractal objects that he called "graftals" (Smith, 1984). Graftal is a formal grammar used in computer graphics to define branching tree and plant shapes in recursively way. The shape is defined by a string of symbols constructed by a graftal grammar. The graftal grammar consists of an alphabet of symbols that can be used in the strings, a set of production rules which translate each symbol into a non-empty string of symbols, and an axiom from which to begin construction. The graftals were built by recursively feeding the axiom through the production rules.

In the same work, Smith coined the term "database amplification" to denote the synthesis of complex images from small data sets.

Later, Prusinkiewicz et al (2007) proposed L+C a modeling language which combined features of L-systems (for example, fast information transfer) and C++. This language extended the L-system formalism and supported a number of standard programming constructs.

Computer graphics connected to the development of new programming languages and to the advanced computer technologies have high potentiality, especially in the field of three-dimensional animations which are connected to the creation of virtual reality environments and virtual worlds.

Virtual worlds are computer-based simulated environments where the users can interact via avatars (our alter ego in the computer representation) in real time with different devices, for example head mounted displays, and data gloves (Damer, 1997).

Avatars can move themselves through virtual territories quickly generated with fractal algorithms which can be codified in different computer languages (Java, C++, VRML). In particular, Virtual Reality Modeling Language (VRML) is the three-dimensional graphics language used on the World Wide Web for producing "virtual worlds" that appear on the display screens using appropriate VRML browsers (for example, Cosmoplayer®, Cortona®).

Figure 9 shows a virtual fractal mountain generated in VRML.

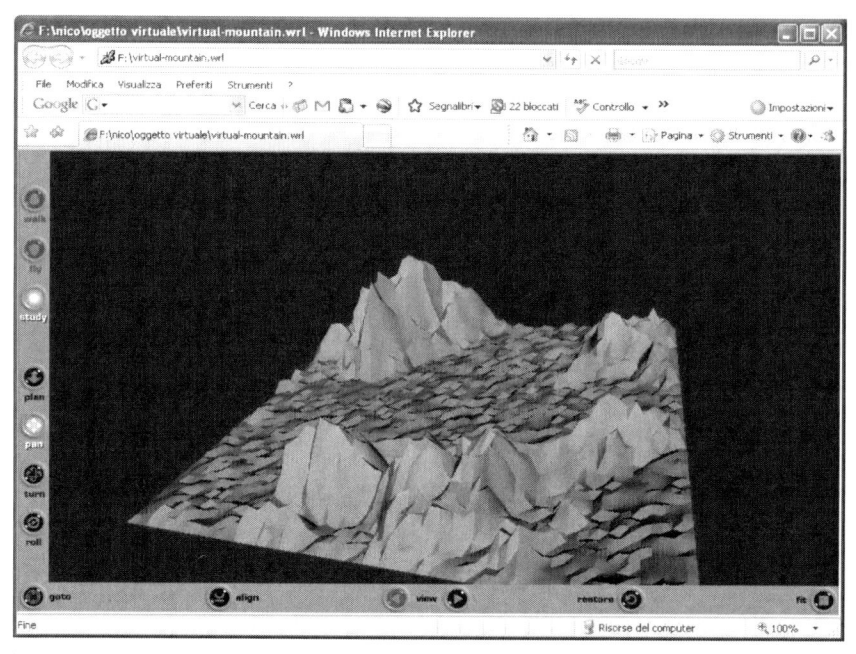

Figure 9. Virtual mountain realized in VRML using simple fractal algorithms.

This example shows that VRML can open new opportunities for using fractal geometry in the Internet based virtual worlds (for example, AlphaWorld and Second Life). AlphaWorld was one of the oldest collaborative virtual world on the Internet that hosted some millions of people from all over the world. Since 1995 AlphaWorld has rapidly grown in size, and now exceeds 60 million virtual objects.

Second Life (SL) is an Internet-based virtual world launched in the summer of 2003. It was developed by Linden Research, Inc (also referred as Linden Lab).

Its users, called "residents", can explore, meet other residents, socialize, participate in individual and group activities, and create and trade items (virtual property).

They can interact with each other through avatars, providing an advanced level of a social network service combined with general aspects of a metaverse.

This is possible using a free downloadable client program called "Second Life Viewer".

Avatars in Second Life are three-dimensional representations of the user. They are controlled and maintained by the user for the purpose of having agency within the virtual environment.

4. FRACTAL GEOMETRY AND SECOND LIFE

In Second Life, fractal geometry is present for creating suggestive landscapes, mountains, trees, sculptures and other fractal shapes using simple algorithms, useful for a quick 3D visualization. Figures 10a and 10b show two different applications of fractal geometry for realizing a fractal tree generated on a base shape, and a "Hilbert Cube" (Segerman, 2005).

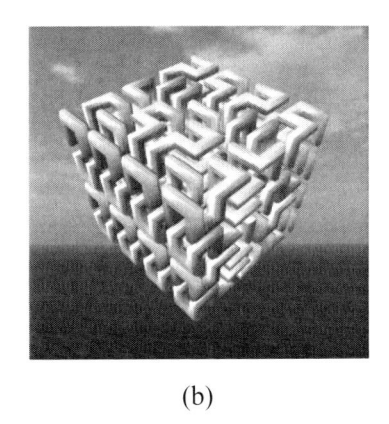

(a)   (b)

Figure 10. (a) Tree automatically generated from a base shape (2005) and copying it in different scales, (b) "Hilbert Cube" (2007) created in 3D using few iterations (http://www.segerman.org/2ndlife.html).

Bourke (2007) presented and interacted in Second Life with 3D geometry derived mathematically or from datasets. The aim of Bourke's work was to explore the representation of various kinds of fractal forms, the applicability to a range of geometric representations was investigated with the view to using Second Life as a way of visualising scientific data in a interactive collaborative environment, which permits remote collaborative exploration of scientific datasets (Bourke, 2008).

In Second Life, other important applications of fractal geometry are in the artistic form production and for virtual art. In this 3D Internet based virtual world, we can find virtual museums which expose fractal paintings and fractal sculptures.

Figure 11 shows an example of fractal objects created in Second life (Bourke, 2007).

Figure 11. Menger cubes in Second Life (Bourke, 2007).

CONCLUSION

This paper described some applications of fractal geometry in computer graphics in particular for modelling complex objects, using Iterated Function Systems, L-systems and the fractional Brownian motion. In the Internet based virtual worlds, fractal geometry can be applied for generating mountains, trees and other shapes, using few iterative or recursive instructions.

Future trends are oriented in different research fields.

One is for using IFS to generate terrain from real data extracted from geological data base. This is useful in the reconstruction of real terrain and landscapes (Guérin et al., 2002).

The other is for integrating Second Life as Internet based virtual world in which one can represent and explore three-dimensional fractals, and in addition, present them to others in a collaborative environment (Bourke, 2009).

REFERENCES

Barnsley, M.F. (1993). *Fractals everywhere*. Boston: Academic Press, 2nd edition.

Barnsley, M.F., Saupe, D., & Vrscay, E.R. (Eds.) (2002). *Fractals in Multimedia*. Berlin, Germany: Springer.

Bourke P. (2007) *Representing and modelling geometry in SecondLife*. Retrieved December 11, 2008, from: http://local.wasp.uwa.edu.au /~pbourke/fractals/secondlife/

Bourke P. (2008) Evaluating Second Life as a tool for collaborative scientific visualization. *Computer Games and Allied Technology*, Singapore, April 28-30, 2008, Retrieved December 14, 2008, from: http://local.wasp.uwa. edu.au /~pbourk/papers/cgat08/paper.pdf

Bourke, P. (2009). Chaos and Graphics: Evaluating Second Life for the collaborative exploration of 3D fractals. *Computer Graphics*, Vol. 33, Issue 3, pp. 113-117.

Damer, B. (1997) *Avatars! Exploring and Building Virtual Worlds on the Internet*. Retrieved February 12, 2007, from http://www.digitalspace.com /avatars/book/chtu/chtu1.htm

Eglash, R. (1999). *African Fractals: Modern Computing and Indigenous Design*. Piscataway, NJ: Rutgers University Press.

Foley, J.D., van Dam, A., Feiner, S.K., & Hughes, J.F. (1997). *Computer Graphics: Principles and Practice. Second Edition in C*. New York: Addison Wesley.

Fournier, A., Fussel, D., & Carpenter, L. (1982). Computer Rendering of Stochastic Models. *Communications of the ACM*, 25, pp. 371-384.

Guérin, E., Tosan, E., & Baskurt, A. (2002). Modeling and Approximation of Fractal Surfaces with Projected IFS Attractors". Novak, M.M. (Ed.). *Emergent Nature: Patterns, Growth and Scaling in the Science* (pp. 293 – 303). New Jersey: World Scientific.

Leland,W.E., Taqqu, M.S., Willinger, W., & Wilson, D.V. (1993). On the Self-Similar Nature of Ethernet Traffic. *Proceedings of the ACM/SIGCOMM'93*, (pp. 183-193) San Francisco, CA.

Lindenmayer, A. (1968). *Mathematical models for cellular interaction in development, Parts I and II.* Journal of Theoretical Biology,*18, pp. 280–315.*

Mandelbrot, B. (1975). *Les Objects Fractals. Forme, Hasard et Dimension,* Paris, France: Nouvelle Bibliothèque Scientifique Flammaron [Fractal Objects. Shape, Hazard and Dimension, Paris, France: New Scientific Library Flammaron].

Mandelbrot, B.B. (1978). Les objets fractals [Fractal objects]. *La Recherche,* 9, pp. 1–13.

Mandelbrot, B.B. (1979). Colliers all´eatoires et une alternative aux promenades aux hasard sans boucle: les cordonnets discrets et fractals [Random necklaces and an alternative to random walks without a loop: the discreet and fractal cords]. *Comptes Rendus* (Paris), 286A, pp. 933– 936.

Mandelbrot, B. (1982).*The Fractal Geometry of Nature.* San Francisco: W.H. Freeman and Company.

Marák, I., Benes, B., & Slavík, P. (1997). Terrain Erosion Model Based on Rewriting of Matrices. *Proceedings of WSCG-97,* vol. 2., pp. 341-351.

Mech, R., & Prusinkiewicz P. (1996). Visual Models of Plants Interacting with Their Environment. *Proceedings of SIGGRAPH 96,* pp.397-410.

Nonnenmacher, T.F., Losa, G.A., Merlini, D., & Weibel, E.R. (Eds.). (1994). *Fractal in Biology and Medicine.* Basel, Switzerland: Birkhauser.

Prusinkiewicz, P., Karwowski, R., & Lane, B. (2007). The L+C plant modelling language. In, J. Vos et al. (eds.), *Functional-Structural Plant Modelling in Crop Production.* (pp. 27-42). Netherlands: Springer, 2007.

Prusinkiewicz, P. & Hammel, M. (1993). A Fractal Model of Mountains with Rivers. *Proceeding of Graphics Interface '93,* pp. 174-180.

Prusinkiewicz, P., & Lindenmayer, A. (1990). *The Algorithmic Beauty of Plants.* New York, US: Springer-Verlag. Retrieved September 10, 2007, from: http://algorithmicbotany.org/papers/abop/abop.pdf.

Sala, N. (2000). Fractal Models In Architecture: A Case Of Study. *Proceedings International Conference on "Mathematics for Living",* Amman, Jordan, November 18-23, pp. 266 – 272.

Sala, N. (2006). Complexity, Fractals, Nature and Industrial Design: Some Connections, Novak, M.M. (Ed.). *Complexus Mundi: Emergent Pattern in Nature.* (pp. 171 – 180). Singapore: World Scientific.

Sala N. (2008). Fractal Geometry in Computer Science, Orsucci F., Sala N. (editors), Reflexing I*nterfaces: The Complex Coevolution of Information Technology Ecosystems*, Information Science Reference, IGI Global, Hershey, New York, pp. 308 – 328.

Segermann, H. (2008). *Second Life*, Retrieved December 18, 2008, from: http://www.segerman.org/2ndlife.html.

Smith A.R. (1984). Plants, Fractals, and Formal Languages, *Computer Graphics*, Vol 18, No 3, July 1984, pp. 1-10 (*SIGGRAPH 84 Conference Proceedings*).

Vyzantiadou, M.A., Avdelas, A.V., & Zafiropoulos, S.(2007). The application of fractal geometry to the design of grid or reticulated shell structures. *Computer-Aided Design*, Volume 39, issue 1, pp. 51-59.

In: Chaos and Complexity in the Arts …　　ISBN: 978-1-53612-995-3
Editors: N. Sala and G. Cappellato　　© 2018 Nova Science Publishers, Inc.

Chapter 9

COMPLEXITY AND FRACTALITY IN INDUSTRIAL DESIGN

*Nicoletta Sala**

Accademia di Architettura Università della Svizzera italiana,
Mendrisio, Switzerland

ABSTRACT

One of the definitions of industrial design could be the following: "Creation and development of concepts and specifications aimed at optimizing the functions, value, and appearance of products, structures, and systems".

Industrial design consists of the ideation of a shape, configuration or composition of pattern or colour. An industrial design can be a two- or three-dimensional pattern used to produce an object. For many years the designers found inspiration by the Euclidean geometry and by the Euclidean shapes (for example, triangles, squares, polygons, Platonic solids, and polyhedra), and it is no surprise that the industrial design's objects have Euclidean aspects. The aim of paper is to present some examples of industrial design objects analysed using the complexity and the fractal geometry. Complex and fractal components appeared in the industrial design after the development of materials, for example the introduction of the float glass, and the manufacturing techniques, for example the work with the laser. This exploration and development of

* Corresponding Author Email: nicoletta.sala@usi.ch.

materials, manufacturing techniques and design are often indistinguishable from one another. This paper is organized as follows: in section 1 there is the introduction. Section 2 describes the complex components in the industrial design. Section 3 presents the fractal components in the industrial design objects. Section 4 describes an application of the box-counting dimension. In section 5 there are the conclusions. Section 6 is dedicated to the references.

Keywords: box-counting dimension, complexity, factuality, fractal geometry, industrial design, self-similarity

1. INTRODUCTION

Industrial design objects could observed to find the presence of mathematical and Euclidean components; for example, the golden ratio, the symmetry, the spirals, the curves, the splines and the surfaces [1, 2, 3, 4, 5, 6]. The evolution of the materials (e.g., glass, plastic, steel, composite), of the technologies (from the hand-made to the Computerized Numeric Control) and of the new approach in the creation processes (e.g., rapid prototyping and hydro forming) have permitted to the designers to overcome the limits imposed by the Euclidean geometry. Thus modern design studies apply complex shapes and fractal geometry to create new kind of objects inspired by the nature or which have futuristic shapes. The objects created in the industrial design can be observed using a different point of view, for example searching complex or fractal components. In our analysis the complexity is not conceived as an area of chaos, but as an attempt to overcome the Euclidean shapes. The complexity is also connected to the fractal geometry that describes the irregular shapes and it can occur in many different places in both Mathematics and elsewhere in Nature. Fractal objects are irregular in shape, and they are generally self-similar and independent of scale. Complexity and fractal geometry can inspire an aesthetic sense [7, 8]. In 1930, the mathematician George Birkhoff (1884-1944) proposed a measure of beauty which involved the "aesthetic measure" (or beauty, M), the order (O) and the complexity (C):

$$M = \frac{O}{C}$$

(1)

This relationship suggests the idea that beauty is connected to order and complexity.

Our complex and fractal analyses in industrial design have been organised in three parts:

1. The complex components in industrial design (e.g., the complex shapes and the complex textures);

2. The fractal component in industrial design objects;

3. The box-counting dimension applied in industrial design objects, to determine their degree of complexity.

2. THE COMPLEX COMPONENTS IN INDUSTRIAL DESIGN

The complexity has generated new kinds of industrial design objects. To research the complex components in this field, we can observe the complex covers (e.g., in the Paul Smith's works) and the complex surfaces (e.g., in the Cini Boeri's monolithic chair realized in float glass, or the Campana's objects created for Alessi) derived by the observation of the nature [6, 8, 9].

Ross Lovegrove designed a PET bottle (1999-2002) for Ty Nant thinking on the water movements (Figure 1a). The bottle is conceptually unique in terms of production delivery and in form to the water concept, demonstrating advances in blow moulding techniques. As the bottle has a fluid asymmetrical form, the surface has and irregular and contoured shape which evokes the fluidity of water (Figure 1b).

Alessandro Mendini conceived, for Alessi, a tea and coffee service (2003) made of in wood carved by hand with inside in 925/1000 silver. We can observe the complex shapes present in the Mendini's project, for example in the shape of the teapot, in Figure 2a, which is similar of the shell with fractal decoration shown in Figure 2b [9].

Alessi's "Tea & coffee towers" project, realized under the control of Mendini, is an example of a new landscape of objects designed by contemporary architects. It realizes an isomorphism between the complexity in architecture (complexity in the large scale) and the complexity in the industrial design (complexity in the small scale) [10]. All designers involved in this project, for example, Zaha Hadid, Massimiliano Fuksas and Doriana O. Mandrelli, Vito Acconci, Tom Kovac, Toyo Ito, Ben van Berkel and Caroline Bos, and others, worked on a dizzy change of scale and on some very different functions [10].

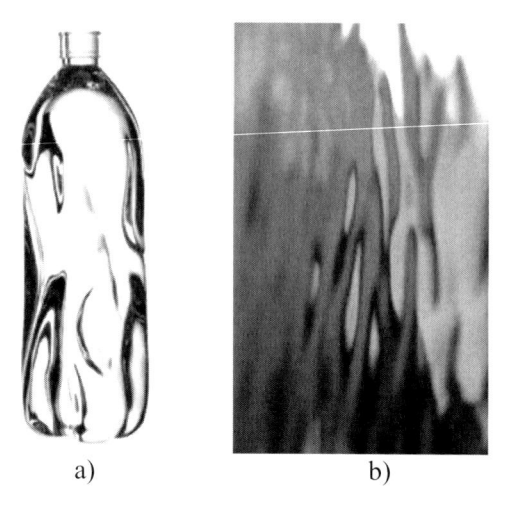

a) b)

Figure 1. (a) TY NANT bottle (2001) by Lovegrove (b) recalls the water motions.

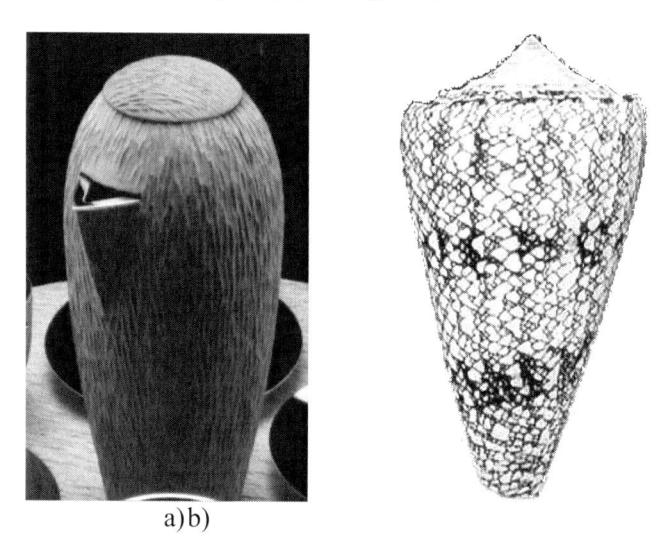

a)b)

Figure 2. (a) Tea and coffee service (2003), realized by Mendini for Alessi (b) shows complex decorations like a shell (Conus araneosus).

Hadid created for Alessi's "Tea & coffee towers" project a coffee-tea pot in 925/1000 silver, in Figure 3, that re-calls futuristic and fractal shapes. She used the "broken" symmetry and the fractal geometry to realize her projects [9]. Observing Figure 3, we note the complexity of the shapes as a 3D puzzles, the single parts form the unity. This object exploits the combination and the contrast between the vertical and the horizontal parts [10].

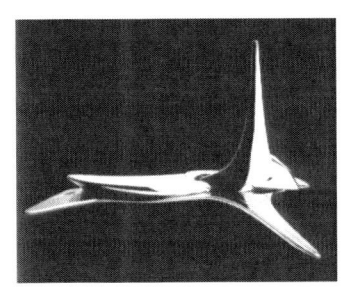

Figure 3. Tea and coffee service (2003), realized by Hadid for Alessi, exploits the combination and the contrast between the vertical and the horizontal parts.

Italian architect and designer Cini Boeri created, in collaboration with Tomu Katayanagi, *Ghost* (1987) a curve float glass monolithic chair, 12 millimetres of tick, for FIAM Italia (Italy). The shape is smoothed and complex and it has realized using a special manufacturing process created by FIAM Italia.

The German lighting designer Ingo Maurer, that raised lighting design and lamps to a high art form, designed the lamp *Paraguadì* (1997), shown in Figure 4, for Guadi's *Casa Botines* (1891-1892). Probably, he found inspiration for this lamp observing the turbulent motions of the smoke.

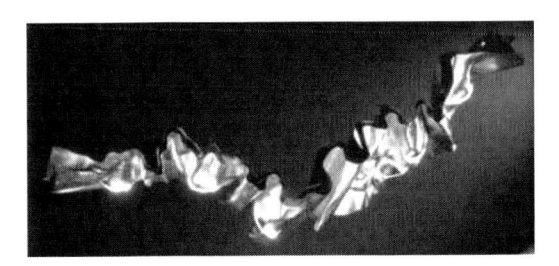

Figure 4. The lamp *Paraguadì* (1997) by Maurer could be inspired to the turbulent motions of the smoke.

Tom Raffield captures the natural complex shapes for realizing everyday *objects. He finds* inspiration from the Cornish environment with his integral passion for designing with wood, Tom marries old techniques with contemporary design to produce his interpretations of objects. *He affirms:* "My ethos is to create products that will be cherished, enjoyed and loved. I also strongly feel, in our disposable culture longevity is the basis for sustainability. In an ideal world, sustainability is a by-product of good design. Running a business in a sustainable way should naturally be sound business practice." *Figure 5a shows a lamp realized by Raffield,* created with 80 metres of steam bent strips of Ash, woven, coiled and twisted around one another, its shape recalls the complex sun activities (Figure 5b).

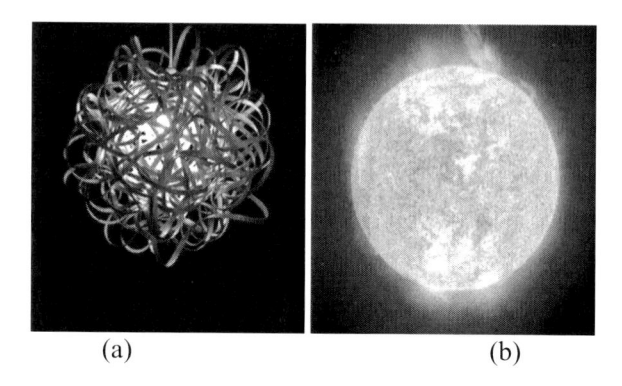

(a)   (b)

Figure 5. (a) *Pendant's No. 1* in ash by *Raffield* (b) recalls the complex sun activities.

Danish architect and designer Poul Henningsen (1894-1967) is widely regarded as a true "master of lamp making." Two of the most prevalent icons of Henningsen's career are the *PH5 lamp*, designed in 1924 and the *PH Artichoke* (1958). The *PH lamp*, also known as the *Paris* lamp for its award winning appearance at the Paris World Exhibition, used tiers of shades, enabling the user to direct light in several different directions without exposing the bulb. The *PH Artichoke*, shown in Figure 6a, is worked on the same principle, although it had even more panels and layers of shades. This Henningsen's object is similar to a real artichoke, in Figure 6b, the vegetable which has inspired the lamp.

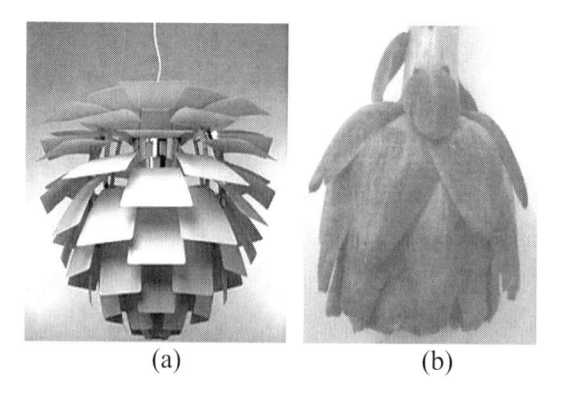

(a)   (b)

Figure 6. (a) *PH Artichoke lamp* (1958) conceived by Henningsen for Poulsen (b) as a artichoke.

The centrepiece bowl *Blow Up* (2004) and the citrus basket *Blow up Bamboo Collection* (2010) realized by Humberto Campana and Fernando Campana for Alessi, shown in Figure 7a and 7b, are other examples of industrial design objects that involve complex shapes that mimic the objects realized using natural materials common in the African and Brazilian cultures. These very complex shapes are realized in 18/10 stainless steel mirror polished and in bamboo wood. *Anemone chair* (2000), Figure 8a, is another realization by Humberto and Fernando Campana for Edra (Italy) which recalls the complex shapes present in the sea (Spirographis Spallanzani) Figure 8b.

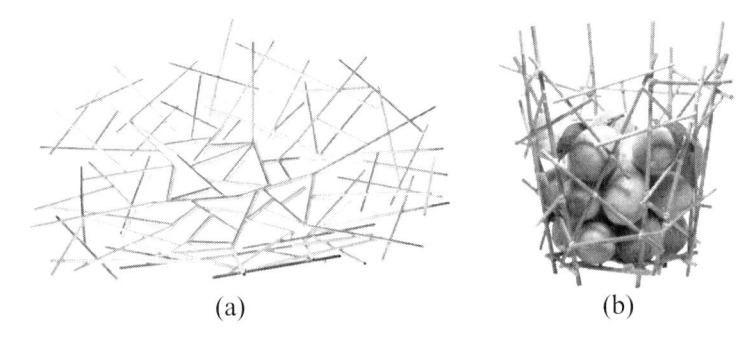

(a)   (b)

Figure 7. (a) Centrepiece bowl *Blow up* (2004) (b) Citrus basket *Blow up Bamboo Collection* (2010) designed by Humberto and Fernando Campana for Alessi.

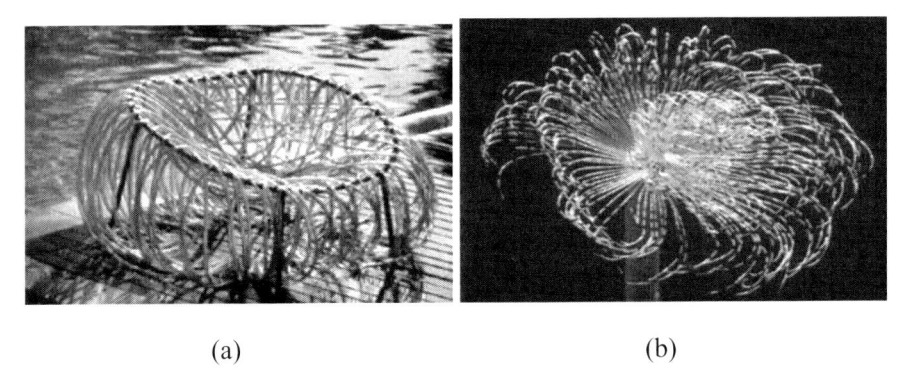

(a)   (b)

Figure 8. (a) *Anemone chair* (2000) designed by Humberto and Fernando Campana for Edra (Italy) (b), recalls the complex shapes present in the sea (Spirographis Spallanzani).

3. THE FRACTAL COMPONENTS IN INDUSTRIAL DESIGN OBJECTS

Fractal geometry describes objects that are scale symmetric, or self-similar. This means that when such objects are magnified, their parts are seen as an exact resemblance to the whole, the property continues with the parts of the parts and so on to infinity. The nature and the characteristics of fractals are reflected in the word itself, coined by the Polish-born French mathematician Benoit B. Mandelbrot (b. 1924) from the Latin verb frangere, "to break", and from the related adjective fractus, "fragmented and irregular" [11, 12, 13, 14, 15].

Irregularity, self-similarity between the original structure and its smaller constitutive fragments, form invariance under changes of measure (scaling) and iteration of unit generator, are main properties which characterize the fractal objects. Mandelbrot used the term "self-similar" for the first time in 1964, in an internal report at IBM, where he was doing research, and in the title of a 1965 paper. A fractal object is self-similar. This means that as viewers peer deeper into the fractal image, we can observe that the shapes seen at one scale are similar to the shapes seen in the detail at another scale. There are three kinds of self-similarity:

- Exact self-similarity. The fractal is identical at different scales. This is the strongest kind of self-similarity.
- Quasi-self-similarity. The fractal is approximately (but not in exact way) identical at different scales. This is a less precise form of self-similarity. Quasi-self-similar fractals contain small copies of the entire fractal in degenerate and distorted shapes. This is the kind of fractals defined by recurrence relations.
- Statistical self-similarity. The fractal has statistical or numerical measures which are preserved across scales; instead of specifying exact scales, at each iteration the scale of each piece is selected randomly from a set range. This is the weakest kind of self-similarity. Most common definitions of "fractal" imply this kind of self-similarity. Random fractals are examples of fractals which are statistically self-similar.

The architect and designer Tom Kovac, Australian of Slovene descent, has realized for Alessi's "Tea & coffee tower" project a tea and coffee service in double wall 925/1000 silver. Kovac is the Creative Design Director of Curvedigital a joint project between RMIT University and Melbourne Museum. He has explored the potential of new processes and developments in the non standard manufacturing. The set of the objects that composes this service is a kind of code which activates a geometric progression, shown in Figure 9. This progression opens new hypothesis in the design's methodologies. The parts that form the set derive by correlation criteria that use the inference by the conceptual to the operative [8, 10]. Observing Figures 9 and 10 the properties of the quasi self-similarity are evident and repeated in four different scales.

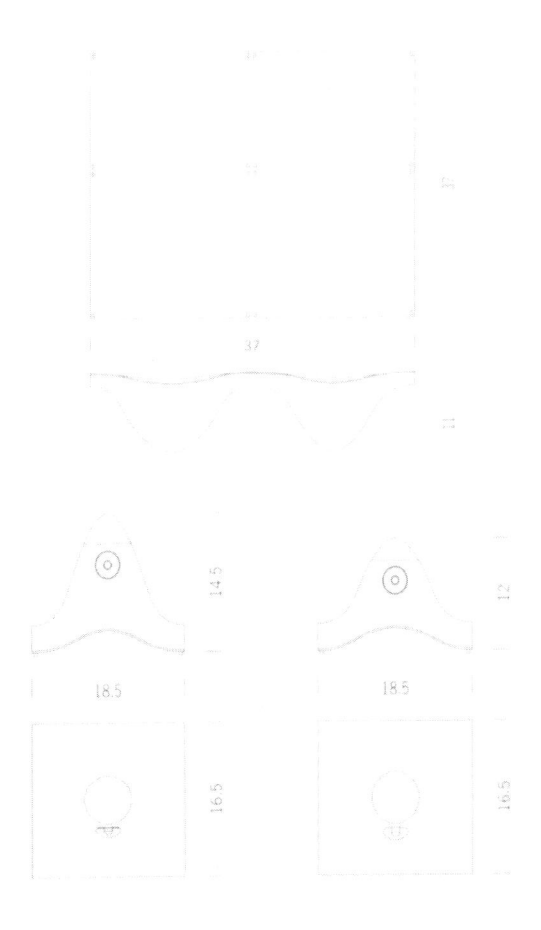

Figure 9. Tea and coffee service (2003) realized by Kovac for Alessi.

4. THE BOX-COUNTING DIMENSION

The box-counting dimension is connected to the problem of determining the fractal dimension of a complex two-dimensional image. It is defined as the exponent D_b in the relationship:

$$N(d) \approx \frac{1}{d^{D_b}}$$

(2)

where $N(d)$ is the number of boxes of linear size d, necessary to cover a data set of points distributed in a two-dimensional plane. The basis of this method is that, for objects that are Euclidean, equation (2) defines their dimension. One needs a number of boxes proportional to $1/d$ to cover a set of points lying on a smooth line, proportional to $1/d^2$ to cover a set of points evenly distributed on a plane, and so on. Applying the logarithms to the equation (2) we obtain: $N(d) \approx -D_b \log(d)$.

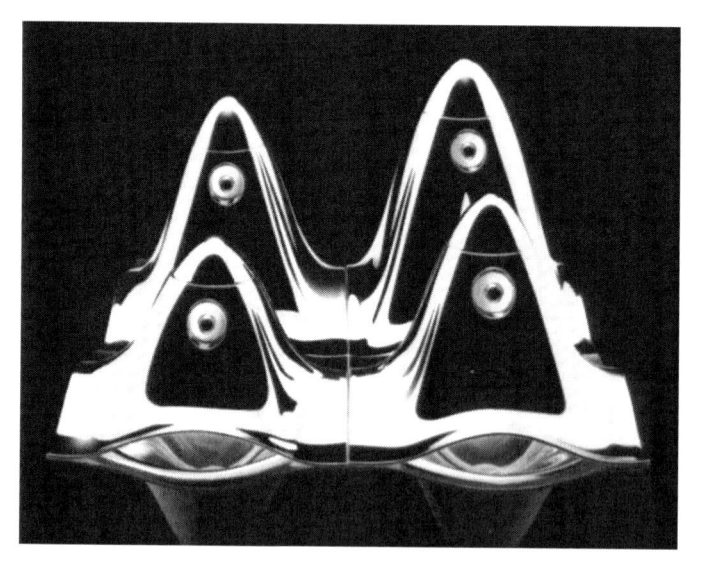

Figure 10. Tea and coffee service (2003) realized by Kovac for Alessi shows the quasi self-similarity applied in four different scales.

The box-counting dimension can be produced using this iterative procedure:

- superimpose a grid of square boxes over the image (the grid size as given as s_1);
- count the number of boxes that contain some of the image ($N(s_1)$);
- repeat this procedure, changing (s_1), to smaller grid size (s_2);
- count the resulting number of boxes that contain the image ($N(s_2)$);
- repeat this procedures changing s to smaller and smaller grid sizes.

The box-counting dimension is defined by:

$$D_b = \frac{[\log(N(s_2)) - \log(N(s_1))]}{\left[\log\left(N\left(\frac{1}{s_2}\right)\right) - \log\left(N\left(\frac{1}{s_1}\right)\right)\right]} \quad (3)$$

where 1/s is the number of boxes across the bottom of the grid. We can apply the box-counting dimension in the industrial design, too. It is calculated by counting the number of boxes that contain lines from the drawing inside them.

To determine the degree of the complexity of the industrial design objects, we applied the iterative methods presented in this section, but we also used the software solutions. The procedure was the following:

- collect the images of industrial design objects;
- digitize the images using a Printer-Copier
- save them in bitmap (*.bmp) format
- analyse the images using a software tool (for example the program Benoit®, version 1.3) in order to calculate Box (D_b), Information (D_i), and Mass dimensions (D_M), and their respective standard errors and intercepts on log-log plots.

Using this procedure we calculated the box-counting dimension of two industrial design objects that evidence a complex shape: the mat *Blow up* (2004), realized by Humberto and Fernando Campana for Alessi and the coffee tables *Papilio* (1985) designed by Mendini in two versions: two and three shelves [8]. Figure 11 shows the box-counting method applied for ten iterations to the coffee tables *Papilio*. Table 1 resumes the results of our analyses.

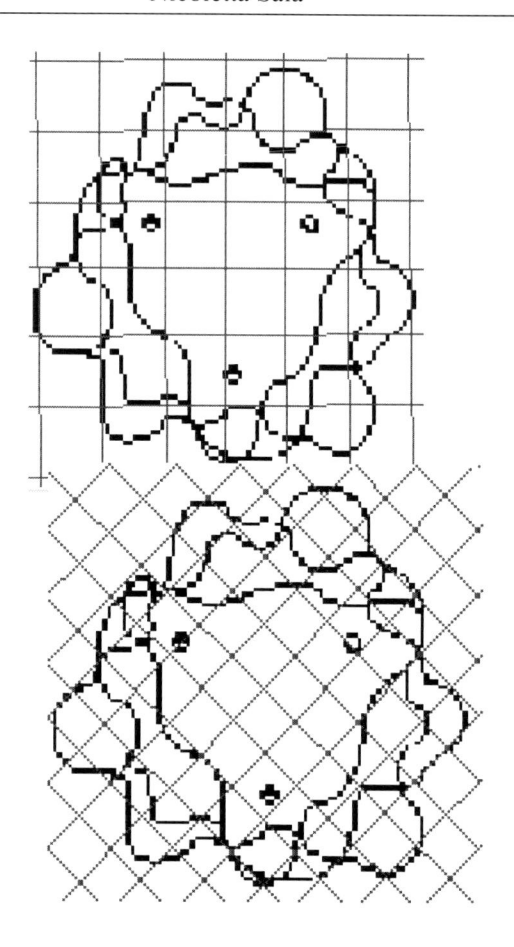

Figure 11. Box counting method applied to the coffee tables *Papilio*.

Table 1. Fractal analyses of industrial design objects

Industrial design object	Box-counting dimension	Information dimension	Mass dimension
Mat *Blow up* (2004)	1.95727±0.0100442	1.93884±0.00071467	1.94470±0.0144529
Papilio (1985) Two shelves	1.88114±0.0079822	1.86924±0.0003498	1.88721±0.0011540
Papilio (1985) Three shelves	1.87092±0.0044575	1.83010±0.0006803	1.87370±0.0012665

CONCLUSION

The complexity paradigm is developing two different traditions: one in the industrial design, another in the science. In this work we described only an approach where the complex and the fractal components have been analysed using three different points of view: to search the complex components in an industrial design object, to determine the self similarity in the industrial design objects, and to find the box-counting dimension applied in the design objects, to determine their degree of complexity. The examples described in the sections 2 and 3 introduce the concept of non-linear design. Non-linear design has influenced Zaha Hadid, Marc Newson, Tom Kovac and other designers.

The evolutions of material, of manufacturing techniques and their mutual exploration, generated great products and "cult" objects of design that have complex shape [16, 17, 18, 19, 20, 21, 22, 23, 24, 25].

In 1931, Le Corbusier affirmed: *"Geometry is the language of man...he has discovered rhythms, rhythms apparent to the eye and clear in their relations with one another. And these rhythms are at very root of human activities..."* [26].

In this century, fractal geometry and complexity could help the man to find new languages in industrial design.

REFERENCES

[1]   Elam, K. *Geometry of design*, Princeton Architectural Press: New York, 2001.

[2]   Fiell, C. and Fiell, P., *Design du XXe siècle* [20th Century Design], Taschen: Köln, 2000.

[3]   Alessi (ed.), *Alessi The Design Factory*, Wiley Academy: New York, 1998.

[4]   Munari, B., *La scoperta del quadrato* [The discovery of the square], Zanichelli: Bologna, 1987.

[5]   Munari, B., *La scoperta del triangolo* [The discovery of the triangle], Zanichelli: Bologna, 1976.

[6]   Alessi, C. (ed.), *Oggetti e progetti. Alessi: storia e futuro di una fabbrica del design italiano* [Objects and projects. Alessi: history and future of an Italian design factory], Mondadori Electa: Milano, 2010.

[7]   Briggs, J., *Fractals the patterns of chaos*, Thames and Hudson: London, 1992.

[8]   Sala, N., Complex And Fractal Components In The Industrial Design, *Int. Journal of Design and Nature.* Vol. 1, No. 2, 2007, pp. 161–173

[9]   Sala, N. and Cappellato, G., *Architetture della complessità* [Architectures of complexity], Franco Angeli: Milano, 2004.

[10] Mendini, A., *Tea and coffee towers*, Electa: Milano, 2003.

[11] Mandelbrot, B.B., *Les objects fractals. Forme, Hasard et Dimension* [Fractal objects. Shape, Chance and Dimension], Flammarion: Paris, 1975.

[12] Peitgen, H.O., Hurgens, H. and Saupe, D., *Chaos and Fractals: New Frontiers of Science*, Springer-Verlag: New York, 1992.

[13] Fivaz, R., *L'ordre et la volupté* [Order and voluptuousness], Press Polytechniques Romandes: Lausanne, 1988.

[14] Sala, N. and Sala, M., *Geometrie del Design: forme e materiali per il progetto* [Design geometries: shapes and materials for the project], Franco Angeli: Milano, 2005.

[15] Sala, N., Fractal Geometry In The Arts: An Overview Across The Different Cultures. Novak M.M. (Eds.) *Thinking in Patterns Fractals and Related Phenomena in Nature*, World-Scientific: Singapore, pp. 177-188, 2004.

[16] Sala, N., Complexity, Fractals, Nature and Industrial Design: Some Connections, M.M. Novak (ed.), *Complexus Mundi: Emergent Pattern in Nature*, World Scientific Singapore, 2006, pp. 171 – 180.

[17] Bangert, A. and Karl, A., *80's Style: Designs of the Decade*, Abbeville Press: New York, 1990.

[18] Bayley, S. (ed.), *The Conran Directory of Design*, Villard Books: New York, 1985.

[19] de Noblet, J. (ed.), *Industrial Design: Reflections of a Century*, Flammarion/APCI: Paris, 1993.

[20] Ashby, M. and Johnson, K., *Material and Design*, Butterworth Heinemann: UK, 2002.

[21] Bejan, A., *Advanced Engineering Thermodynamics*, second edition, Wiley: New York, 1997.

[22] Bejan, A., *Shape and Structure, from Engineering to Nature*, Cambridge University Press: Cambridge, UK, 2000.

[23] Hernandez, G., Allen, J.K. and Mistree, F., Design of hierarchic platforms for customizable products, ASME Paper DECT2002/DAC-34095, *Proceedings of DETC'02, ASME 2002 Design Engineering Technical Conferences and Computer and Information in Engineering Conference*, Montreal, Canada, September 29-October 2, 2002.

[24] Carone, M.J., Williams, C. B., Allen, J.K. and Mistree, F., An application of constructural theory in multi-objective design of product platforms, ASME Paper DECT2003/DTM-48667, *Proceedings of DETC'03, ASME 2003 Design Engineering Technical Conferences and Computer and Information in Engineering Conference*, Chicago, Illinois, September 2-6, 2003.

[25] Rosa, R.N., Reis, A.H. and Miguel, A.F., *Bejan's Constructal Theory of Shape and Structure*, Évora Geophysics Center, University of Évora, Portugal, 2004.

[26] Le Corbusier, *Towards A New Architecture*, Dover Publications Inc: Mineola, NY, 1985.

In: Chaos and Complexity in the Arts …　　ISBN: 978-1-53612-995-3
Editors: N. Sala and G. Cappellato

Chapter 10

ARCHITECTURE AND TIME

Nicoletta Sala[*]

Accademia di Architettura, Università della Svizzera italiana
University of Lugano, Mendrisio, Switzerland

«Architecture is the masterly, correct, and magnificent play of masses brought together in light.»

Le Corbusier (1887-1965)

ABSTRACT

Architecture is a discipline which involves geometry, mathematics, physics and engineering. The time is a parameter which architects seldom consider in their projects.

The aim of this paper is to show how, and where, the concept of time could be applied in architecture.

Keywords: architecture, dynamic architecture, dynamic façades, geometry, rotating skyscraper, time

[*] corresponding Author Email: nicoletta.sala@usi.ch.

1. INTRODUCTION

Architecture is a language and it is the summary of the major arts, in fact it involves form, volume, colour, acoustics, mathematics, music and engineering [11, 12].

Time is a word derived by the Latin substantive *tempus*, that more properly means "duration fraction" or "time moment"; only later it has assumed the meaning of "every duration fraction".

Encyclopaedia Britannica defines the time as (2011): «a measured or measurable period, a continuum that lacks spatial dimensions. Time is of philosophical interest and is also the subject of mathematical and scientific investigation» [24].

Two different points of view on time divide many philosophers. One is oriented to consider the time as a part of the fundamental structure of the universe, a dimension where the events occur in sequence. For example, Newton applied this realist view, known as Newtonian time.

Second viewpoint is to consider that the time does not refer to any kind of actually existing dimension that events and objects "move through", but it is an intellectual concept that enables humans to sequence and compare events.

Kant presented, in the Transcendental Aesthetic of his *Critique of Pure Reason*, a series of arguments and conclusions concerning the nature of time. His basic idea was to demonstrate that time is presupposed in all human experience, although not for the ordinary reason that it is in some way inherent in the nature of the universe apart from consciousness [22].

He wrote (1781): «Time is not an empirical concept that is somehow drawn from experience. For simultaneity or succession would not themselves come into perception if the representation of time did not ground them a priori. Only under its presuppositions can one represent that several things exist at one and the same time (simultaneously) or in different times (successively)» [9].

In the last century, Einstein postulated that space and time are into a single continuum (called: spacetime, or space–time, space time).

Architecture involves many disciplines. For example, engineering, for structural calculi, Euclidean and non-Euclidean geometries, to reorganize the space and to study the urban growth, and physical techniques to solve energetic problems [2, 11, 12, 13, 14, 15 16, 17, 18, 19, 20, 21, 26].

This paper, that describes how the time could be applied in architecture, has organized as follows: section 2 analyses the connections between architecture and time. For example, rotating buildings, dynamical façades and rotating skyscrapers. In section 3 there are the conclusions, and section 4 is dedicated to the references.

2. ARCHITECTURE AND TIME

Few architects have considered the role which the time has on their buildings. For example, the time and the light could break the symmetry of an architectonic design, creating sensations of caducity, or new ties which the architect has not thought [18].

Taj Mahal (Agra, India, 1632-1648,) is and example of Moghul art, and its name is the deformation of "Muntaz Mahal" which means "the favourite of the Harem". In this mausoleum the light creates on the water different effects and symmetries, in two different moments of the day, as shown in Figure 1.

Francesco Venezia affirms: «We draw only a half of the building, the immutable one. The second, half of shades, forms and changes itself in every time of the day in every day of the year. The architecture glamour is in expecting this change through the project, through the geometry. But a little cloud is enough to cancel all what we have expected. And that gives to the architecture one "moving" path in the days and seasons light. The same one for the reflex. Paul Valery writes: "O mon semblable et pourtant plus parfait que moi-même". It is important in architecture to evaluate the exact role of the reflex image. We can double the bodies and the double can be more perfect than the real one. But a sudden breeze can cancel this image.»[1] [26].

The time and the atmospheric agents corrode and damage the buildings modifying many architectures, and the nature has worked on them, creating new hybrid forms. Previous Figure 2 shows some examples in Angkor Wat (Cambodia), where the nature worked on the temples, creating new shapes. Angkor Wat is a temple complex, built in the early 12th century by

---

[1] «Noi disegniamo solo una metà dell'edificio, quella immutabile. La seconda, la metà d'ombre, si forma e si trasforma in ogni ora del giorno in ogni giorno dell'anno. Il fascino dell'architettura risiede nel prevedere questo cambiamento attraverso il progetto, attraverso la geometria. Ma basta una piccola nuvola per cancellare tutto ciò che abbiamo previsto. E ciò dà all'architettura una via "patetica" nella luce dei giorni e delle stagioni. Lo stesso per il riflesso. Paul Valery ha scritto: "O mon semblable et pourtant plus parfait que moi-même". In architettura è importante valutare il ruolo esatto dell'immagine riflessa. Possiamo raddoppiare corpi e il doppio può essere più perfetto del reale. Ma una brezza improvvisa basta a cancellare questa immagine.» [26, p. 82]

Suryavarman II, king of the Khmer Empire, as his state temple and capital city.

Figure 1. *Taj Mahal* (Agra, India, 1632-1648,) the light and the water create different symmetries in different moments of the day.

Figure 2. *Angkor Wat* (Angkor, Cambodia). Going by the time corrodes the architecture and the Nature worked on them. Tree roots are tight welded to the walls of temples, forming a unique architectural structure.

How can we insert the time in modern architecture?

Frank O. Gehry proposes an approach to a dynamic architecture, breaking the symmetry in his buildings. His *Rasin Building* (1992-1995), realized in cooperation with Vlado Milunic, is an example in this field.

It reflects a woman and man (Ginger Rogers and Fred Astair) that are dancing together, for this reason is dubbed *Ginger and Fred*, as the famous Hollywoodian couple, Figure 3a [6].

The site, where Gehry's building has been located, was originally occupied by a Neo-renaissance house from the end of the 19th century until 1945 (year in which the house was destroyed during a bombing).

The *Rasin Building* marks the corner at the *Jiraskuv Bridge* (Prague, Czech Republic). It is an example of deconstructivist architecture, with an unusual shape, Figure 3b.

An other contribution to underscore the dynamism of the construction techniques has been done by Italian architect Paolo Portoghesi, that gives the volumes plasticity and enriching the spatiality and the reflection of the light on their surfaces [14]. Portoghesi finds inspiration by the chaotic movements connected to the gas and liquid motions, to realize *Hotel Savoia* (1992-1996) in Rimini (Italy), particulars of the model in Figure 4a. The building contains smoothed surfaces inspired by the wave motion as a metaphor of the sea, Figure 4b [19, 20]. The hotel is situated on the Adriatic [14].

Dynamism in Portoghesi and Gehry designs is only an attempt to insert the movement in the buildings, breaking the immobility, connected to the Euclidean geometry.

Figure 3. *Rasin Building* (1992-1995), by Gehry and Milunic, reflects Ginger Rogers and Fred Astaire that are dancing together.

Figure 4. *Hotel Savoia* (Rimini, Italy, 1992-1996), by Portoghesi, has been inspired by the waves of the Adriatic sea.

An other attempt to consider the time in architecture is modifying the position which the building has in different moment of the day, for following the sun.

An interesting example is located in Italy. *Villa Il Girasole*[2] (1929-1935) (*Villa Sunflower*) is the only one and well preserved modern and revolving habitation. It is placed in Marcellise, in the countryside of Verona (Italy). It is a revolutionary reinforced concrete architecture, conceived and built by Angelo Invernizzi (1884-1958), an Italian engineer. He thought on the sun's path, on a relation with landscape and the space of human's life.

Villa Girasole's volumes (5000 cubic meters), L-shaped, rotate around a cylindrical vertical board placed on a suitable support platform (44,50 meters in diameter) [8]

In the middle of the building there is a turret (42 meters tall), a sort of conning tower or lighthouse, which the rotating movement hinges on.

The movement is guarantee by a diesel engine which pushes the house over three circular tracks where 15 trolleys can slide the building (the speed is 4 millimeters per second, and it takes 9 hours and 20 minutes to rotate fully). *Villa Girasole* is like a real sunflower which rotates following the sun light.

The time could be inserted in the architectural projects in the building façades modification. The role of the façades in the modern buildings is radically changed in the last few years: they are not only a covering of protection from the atmospheric agents. They also become an element of communication, of accumulation and supply of energy and light, of regulation of thermal and acoustic comfort.

---

[2] Academy of Architecture Mendrisio (USI, Switzerland) actively participates to the preservation of the building, with the "Fondazione Il Girasole" (Sunflower Foundation), created on 11th of April, 2002, with a donation by Lidia Invernizzi, daughter of the house creator.

Figure 5. *Villa Girasole* (Marcellise, Verona, Italy, 1929-1935). The wheels are visible to facilitate necessary repairs.

Toyo Ito conceives the façades as media for sharing information; his *The tower of the wind* (Yokohama, 1986) is an example in this field.

Ito used an old water tower and ventilation conducts in front of the Yokohama bus station to create a concrete structure. He proposed to cover the tower with acrylic mirrors, using a metallic oval-cylindrical structure around the tower (21 meters in height and 9 x 6 meters section), lined with a perforated aluminum coating that reflected the sky during the day (in Figure 6).

The fluxes of the wind are under control by an electronic system which recognizes the differences in the wind speed and in the noise produced by urban traffic translating them in electric impulses able to animate the skeleton (made of steel) using lights and colours in constant change. For this reason, unlike a traditional arrangement of lights, *Tower of the winds* does not follow a computer program, but it offers an answer, as light and color, to the different real dynamic solicitations of the winds and of the noises.

Ten years later, this extraordinary Ito's idea, that was projecting the surrounding environment in a variegated luminous stimulus kaleidoscope, has stopped and it has been overwhelmed from the carelessness and the lack of maintenance.

Egg of the winds (Okawabata River City, 1988-1991), conceived by Ito, is another example where the façades communicate information. It is an ellipsoid (16x8 meters) covered with a perforated aluminum panels which reflect the town images; below a lot of liquid crystal screens display news and images. This capsule is a kind of video gallery outdoor, and it creates very interesting visual effects. The images are floating on a curved surface, and it is different from the visual effect of the giant screens hung on the façades of buildings (Figure 7). [15]

Figure 6. *The tower of the wind* (Yokohama, 1986) by Ito. It translates the noise, produced by urban traffic, in electric impulses able to animate the skeleton through lights and colours in constant change.

About the different use of the façades Tzempelikos affirms (2007): «Each building requires a different design approach, depending on the type of use, climate, orientation and transparency. During the early design stage, the building design team has to choose from a wide variety of design options, for many of which the evaluation of their impact on building performance could be difficult, especially for innovative technologies. Inevitably, the selection of final design solutions often involves many subjective factors. The fragmented nature of the building process, in which no member of the design team considers the overall optimization of the indoor environment, further compounds the problem. Therefore, traditional passive designs are often suggested as the "safe" traditional solution in the final stage» [25].

Dynamic exterior of *Kiefer Technic Showroom* (2007) in Bad Gleichenberg (Austria) consists of electric window shutters made of perforated aluminium. It is designed by Ernst Giselbrecht (+ Partner ZT GmbH), and it changes throughout the day, transforming the building into a dynamic sculpture, as shown in Figure 8. It represents a unique and "Dynamic Façade" [1].

Other example is FLARE, conceived by Staab Architects. It is a modular system which creates dynamic façades for any wall typologies or buildings. The system consists of a number of metal flake bodies supplemented by individually controllable pneumatic cylinders, and it offers one of the more sophisticated electromechanical systems. It is also acclaimed for its ability to adapt to a variety of surfaces, whether it be. [10, 23]

Figure 7. *Egg of the winds* (Okawabata River City, 1988-1991) by Ito. It is a kind of video gallery outdoor, and it creates very interesting visual effects.

Figure 8. *Kiefer Technic Showroom* (Bad Gleichenberg, Austria, 2007) has a "Dynamic Façade".

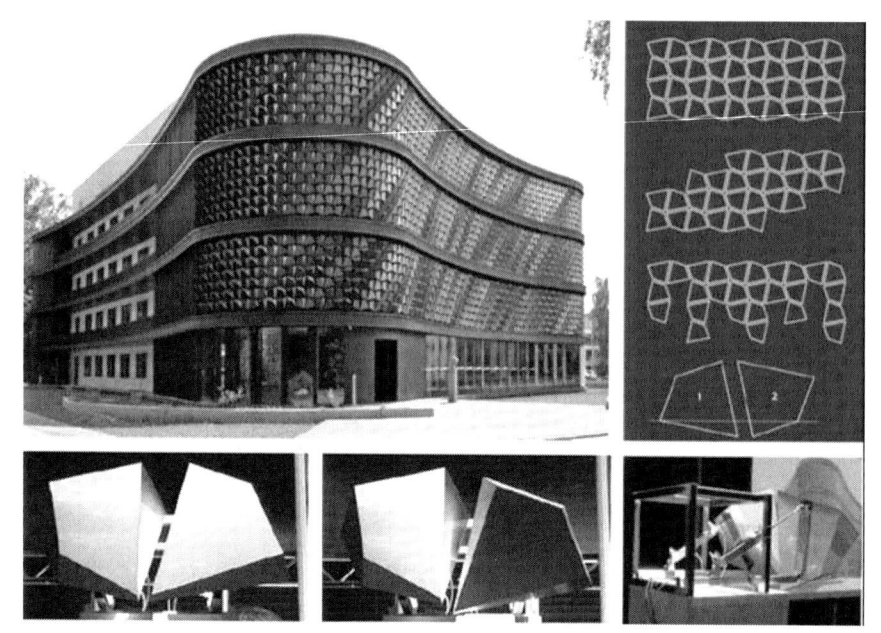

Figure 9. *FLARE system* can create geometric patterns, with an infinite array of flakes which can be mounted on any wall surface.

FLARE system, shown in Figure 9, can create geometric tessellations and patterns, with an infinite array of flakes which can be mounted on any wall surface. Every metal flake can reflect the «...bright sky or sunlight when in vertical standby position. When a flake is tilted downwards by a computer controlled pneumatic piston, its face is shaded from the sky light and appears as a dark pixel. By reflecting ambient or direct sunlight, the individual flakes of the FLARE system act like pixels formed by natural light.» [7, 23].

David Fisher designed *Dynamic Architecture Building* (2011, Dubai, United Arab Emirates). It is also known as *Da Vinci Tower*, or *Dynamic Tower*. It is a moving skyscraper, 420-meters (1,378 ft) 80-floors, that could open new opportunities for the future architecture. Fisher's dynamic skyscraper has introduced three revolutionary aspects in contemporary architecture [3, 4, 5]. First aspect is in the shape of the building, which changes continuously following, for example, the sun and the wind. Each floor can rotate in independent way by the others, changing the shape of the tower. For example, we can get up with the sun rising into our bedroom and enjoy the sunset over the ocean at evening.

Fisher describes *Da Vinci Tower* as «planned from the Life, forged from the time» [5].

Figure 10. *Da Vinci Tower* (2011, Dubai, United Arab Emirates). Each floor can rotate a maximum of 6 meters (20 ft) per minute, or one full rotation in 90 minutes.

Building's shape is imposed by time and by life demands, and it can be fitted to the demands of the flat owners, the living room following the sun movement or the bedrooms oriented to north-south. This rotating tower is the first building which has four dimensions: three co-ordinates for the space and one for the time. The second innovative aspect is connected to the method of construction: prefabrication. The entire building is made of prefabricated modules (aluminum, steel, carbon fibre and other modern materials) produced in factory which arrived to the construction yard. They were installed together in a very short construction time, with a limited number of workers, reducing the costs and the risks of accidents on the site. The skyscraper, made of single parts connected together, has a high seismic resistance.

The third innovative aspect is the combination between technology and environment. The building, in continuous movement, is energetically autonomous. It has forty-eight wind turbines, positioned horizontally between each floor, and many photovoltaic panels which produce "green" energy able to furnish power to the neighbouring buildings. In this way, the skyscraper, self-powered, becomes a "power station". Previous Figure 10 shows some particulars of *Da Vinci Tower*.

CONCLUSION

The time is a parameter which architects seldom consider in their projects.

In contemporary architecture the dynamic façades, changing their shapes, allow a building to communicate and interact with their environment.

Dynamic building envelopes include advanced window technologies, innovative fenestration systems and automated shading control, all of which characterize the new "intelligent" buildings generation. Although a great idea, the design and implementation of such systems is a very complex task.

Fisher's Dynamic Skyscraper has introduced three innovative aspects in contemporary architecture. In particular, this skyscraper becomes a "power station" for the city.

Rotating towers, changing shape, will end the era of the static and immutable architecture, marking the beginning of a new one, where the keywords will be: dynamism and adaptability to the life demands.

This will result in a constant modification of the tower's shape. Each floor will rotate a maximum of 6 metres (20 ft) per minute, or one full rotation in 90 minutes.

Fisher's dynamic buildings could become a symbol of a new philosophy in the construction techniques which will change the image of the cities, and they could realize a new connection between the time and the architecture.

REFERENCES

[1] *Architectural Design of Dynamic Facade by Giselbrecht + Partner ZT GmbH* (2009). Retrieved 15 April 2011: http://archometrend.blogspot.com/2009/11/architectural-design-of-dynamic-facade.html

[2] Batty, M., Longley, P. (1994). *Fractal Cities: A Geometry of Form and Function*. Academic Press, San Diego, CA and London.

[3]   *Dynamic Architecture*. Retrieved 10 February 2011: http://www.
      dynamicarchitecture.net/index.php?option=com_contentandview=article
      andid=37andItemid=10andlang=eng

[4]   Hope, B. (2009). *Towers take turn for the worse*. The National, 29 June
      2009. Retrieved 1 February 2011: http://www.thenational.ae/business/
      property/towers-take-turn-for-the-worse.

[5]   Rotating Tower - *Dynamic Architecture: The Dynamic Architecture*
      Retrieved 15 April 2011: http://www.gruppogedi.it/the_dynamic_
      architecture.htm.

[6]   Dal Co, F, Forster, K., (2003). *Frank O. Gehry: The Complete Works*.
      Phaidon Press, London.

[7]   *Flare-façade* http://www.flare-facade.com/.

[8]   Frampton, K, Galfetti, A. (2006). *Villa Girasole: The Revolving House*.
      Mendrisio Academy Press, Mendrisio, Switzerland.

[9]   Kant, I. (1781). *Critique of Pure Reason*. Trans. Norman Kemp Smith
      with preface by Howard Caygill. Pub: Palgrave Macmillan. Retrieved,
      20 December, 2010: http://www.hkbu.edu.hk/~ppp/cpr/toc.html.

[10] *Kiefer Technic Showroom: Architecture Information*. Retrieved 10 April
      2011:   http://www.e-architect.co.uk/austria/kiefer_technic_showroom
      .htm.

[11] Le Corbusier (1923). *Vers une Architecture* [Towards Architecture].
      Crès, Paris.

[12] Le Corbusier (1985). *Towards a New Architecture*. Dover Publications,
      New York.

[13] Moore, R. (2011). *Like Spinning Plates*. Specifier. Retrieved, 6 April
      2011:   http://www.specifier.com.au/currentissue/45752/Like-Spinning-
      Plates.html.

[14] Portoghesi, P. (1999). *Natura e Architettura*. Skira, Milano (English
      version: Nature and Architecture, Skira, Milano, 2000).

[15] Prestinenza Puglisi, L. (1999). *Hyper architecture: spaces in the
      electronic age*. Birkhäuser, Basel.

[16] Rocca, F.X. (2009). Believe Him or Not, He Puts a Fresh Spin on
      Architecture. *WSJ.com*. Retrieved 3 February, 2011: http://online.wsj
      .com/article/SB123432213609971519.html.

[17] Sala, N., Cappellato, G. (2003). *Viaggio Matematico Nell'arte e
      nell'architettura* [Mathematical Journey in Art and Architecture]. Franco
      Angeli, Milano.

[18] Sala, N., Cappellato, G. (2004). *Architetture della complessità. La
      geometria frattale tra arte, architettura e territorio* [Architectures of

complexity. Fractal geometry between art, architecture and territory]. Franco Angeli, Milano.

[19] Sala N. (2004). Complexity In Architecture: A Small Scale Analysis. *Design and Nature: Comparing Design in Nature with Science and Engineering.* Wit Press, 2004, pp. 35-44.

[20] Sala, N. (eds.) (2008). *Chaos and Complexity in Arts and Architecture.* Nova Science, New York.

[21] Salingaros, N.A. (2005). *Principles of Urban Structure.* Techne Press, Delf.

[22] Smith, N.K., Gardner S. (2003). *A Commentary to Kant's Critique of Pure Reason* (2nd edition). Palgrave Macmillan, London.

[23] *SOM + SCI-Arc on CF: Responsive Kinetic Facade* (2009) http://www.core.form-ula.com/2009/04/15/som-sci-arc-on-cfresponsive-kinetic-facade/.

[24] Time. (2011). In *Encyclopædia Britannica.* Retrieved from http://www.britannica.com/EBchecked/topic/596034/time.

[25] Tzempelikos, T. (2007). *Integration of Dynamic Façades with other Building Systems.* Retrieved, 10 April 2011: http://www. automated buildings.com/news/may07/articles/concordia/070428064606conc.htm.

[26] Venezia, F. (1990). *Scritti Brevi*, Clean, Napoli.

In: Chaos and Complexity in the Arts …
ISBN: 978-1-53612-995-3
Editors: N. Sala and G. Cappellato

Chapter 11

SYNCHRONIZATION OF FRACTALS IN LOGARITHMIC SPIRALS

Robert A. M. Gregson[*]

Australian National University, Canberra, ACT, Australia

ABSTRACT

Synchronization is not treated here is a fundamental necessary property, but is transient and derivative on sequences that are generated by nonlinearity in time series which arise in neural brain processes, and in bottom-up top-down network dynamics.

We illustrate in figures some properties using Markov matrices, open symbolic dynamic nets, and fields on Julia sets. The work by Vrobel (2011), in Chapter 6, about Temporal Binding: Synchronizing Perceptions, is cited. We find both symmetrical and spiral patterns on local regions of Julia sets, and discontinuous series in the dynamics of some region that are recordable in the neurophysiology of intermittent consciousness. Synchronization can also be called self-similarity, in induced noncommutative geometry.

Keywords: logarithmic spirals, synchronization, neural networks, Julia sets, fractals

---

[*] Corresponding Author: Email: ramgdd@bigpond.com; Address: Research School of Psychology, Australian National University, Canberra, ACT, Australia.

INTRODUCTION

The problem we are addressing here involves tracing across three stages of dynamic information processing in mental activity, the first statements are presented as time series equations, then secondly that can be read for some bottom-level stages into a chaotic neural network (CNN), then thirdly as down-loading to a computer graphic display. The stages are presented in Equations 1 and 2 and Figures 1 to 4 in that sequence.

In processing the second stage there are two approaches, either to remain continuously in computer algebra and only come to visible printable images after the Julia set dynamics are displayed, as pioneered by Mandelbrot (2004), the other is to put the time series equations into neural network regions, such as EEG or fMRI, and then come back to printing out images of brain activity in selected regions.

This second stage does not necessarily create conscious experience, though under some limited conditions it can transiently do that by meeting conditions called binding. The neural network in based in a brain volume with about 10^{12} elements, and pairs of elements inter-connected by rapidly over as few as four internet cellular axon connections. Obviously we cannot model such a vast set of cells and axons, so a reduced brain structure is postulated, with specialized areas such as frontal or temporal regions.

The third final output involved fractal patterns that include spirals singly or in symmetric logistic spiral pairs, or at different scale magnitudes.

BINDING AND SYNCHRONIZATION

Vrobel (2011, p.118-119) summarized very pertinently that "The *binding problem* of perception addresses the question of how we are capable of combining, with a window of approximately 100-200 milliseconds, such various features of color, texture, smell, auditory or tactile stimuli into a representation of an object.. One possible explanation "(is that)" Gamma waves of 40 Hz. are ubiquitous in the human brain…they phase lock… and trigger temporally synchronous discharge in separate parts of the cortex."

A binding process has been offered as a necessary and sufficient condition for the emergence of transient consciousness (Singer, 2001, Crick and Koch, 2003). The original work on binding was almost fully based on visual processes, (Treisman, 1998, p.1296) commented "binding might depend on synchronized neural activity, but synchronization is much wider."

Synchronization simply means that all oscillations in a neural network eventually phase-lock, but not necessarily in precise timing when the connections are between two human subjects and not within one brain or organism. It is important not to over- invoke the possibility of the brain being functioning as synchronization, because neural avalanches, oscillating and synchronization are three different processes in network states (Beggs and Plenz, 2003; Kozma and Freeman, 2008).

Engel et al (1999, p. 146) concluded cautiously "that an enhancement of synchronization leading to a selection of neuronal populations may be based on a combination of both "bottom-up" and "top-down" influences". It appears that synchronization follows from binding, but binding does not necessarily imply full synchronization. Singer and Gray (1995) showed that presence of stimulus-dependent synchrony between units in quite widely separated areas of the brain that exists.

In order to have an accurate algebraic model to identify the dynamics involving fractals, it is expedient to construct a map of the relevant Julia set. This can then be explored at local regions, and by progressively reducing scales and rotations.

This work was first published by Mandelbrot for a complex quadratic recursion,

$$Y_{j+1} = Y_j^2 + Y_c \quad Y_0 = 0 \tag{1}$$

The derived fractal work is illustrated in the extensive book by Becker & Dörfler (1988). The $\{Yj\}$ are typically expressed in voltages, as found in EEG pathways; they can be in real or complex measured time series, and in plots of resulting bifurcations. The other examples that created such fractal work we refer to here were by Gregson (1992, 1995, 2006) and employed a cubic complex polynomial recursive psychophysical time series (with environment constants $a,e,$) called Gamma:

$$\Gamma_{\text{def}}: Y_{j+1} = -a\,(Y_j- 1)(Y_j + ie)(Y_j - ie),\ i^2 = -1 \tag{2}$$

The Julia set for Gamma is shown enlarged on the front colored book cover on Gregson (2006). The later extension to considerations of synchronized local series was mainly due to Vrobel (2011), who again used the Mandelbrot figure with surrounding extended fields.

THREE WAYS OF REPRESENTING DYNAMICS IN BRAIN ACTIVITY

The literature on consciousness is so extensive that it would need a series of volumes to be summarized, but the introduction of neurophysiological recording of parts of human and animal brain activity, using for example fMRI, EEG and TMS as standard practice in experimental psychology, replacing philosophical speculation, makes it now possible to essay a review of three related approaches, as anticipated and listed in the Introduction.

(*a*) Distinctions within levels of consciousness and unconsciousness, such as attention, waking, sleep, memory, dreaming, and coma.

(*b*) Top-down bottom-up loop flows in nonlinear dynamics,

(*c*) Fractal Julia maps of psychophysical time series, in the major specific senses such as vision, hearing, taste and touch.

This creates a tentative attempt to bring these three (***a,b,c***) together, because each is in itself necessary but conspicuously incomplete.

Vrobel defined (2011, p.17) three states of classification between:

- Δ_t length, during relations not in simultaneity, such as psychophysics in series, that is in succession,
- Δ_t depth during compatible events in simultaneity, nested simultaneous rhythms require more than one dimension and are therefore temporal fractals plotted in 2-d, and
- Δ_t density as a fractal dimension of a time series. We may consider that in (***c***).

(*a*) This is a conjecture influenced by the Seth Suzuki and Crichley (2012) theoretical map of an interoceptive predictive coding. The default mode network (dmn) is postulated to exist in two forms, I have put them in a middle region over the S series within the t, t+1 time transition matrix.

It is important to note that the terminology used by Nicolis and Basios (2015, p. viii)calls Figure 1 the *syntactical* level where the elementary dynamic processing are taking place, while Figure 2 is called the *semantic* level, where relationships between stimuli and the formation of categories are being shaped. The two are both necessary and require to be distinguished in dynamic informational theory.

t+1

	1	2	3	4	5	6	7	8	9
1	S	T	C	NC					
2	B	S	T		NC				
3		B	dmn	S			AG	PR	
4			B	dmn	T			AG	PR
5				B	dmn	S			
6					B	S	T		
7						B	S	T	
8	sensory :						B	S	T
9	motor body:							B	S

Figure 1. One-step Transition Probabilities between a closed set of states.

This is only an outline of the symbolic dynamics of the 9x9 Markov matrix, where row 1 is at top and row 9 is at bottom, and if the distinction is expanded for, say, the three states of awake, asleep and in a coma, to a 3 x 3 higher Markov into which the symbolic matrix shown above is embedded, then the one-step shift from t to t+1 is not stationary.

Conscious and non-conscious (C,NC) must be in top-down regions. In simplest state probability transition matrices, if the leading diagonal top-left to bottom-right cells are empty and reserved for local stationarity (S), then the upper right triangular sub-matrix is top-down (T) and the lower left triangular region is bottom-up (B). Agency (AG) and Presence (PR) are weak top-down locations within middle neural networks.

This neural network matrix has within it a subset of cells with local singularity dynamics, that are not stable over time.

A matrix without any numeral quantification is by definition a form of symbolic dynamics. The two Figs. 1 (above) and 2 (below) respectively exemplify the distinction between structure and function in quantity (Tufte, 2001/2011). Real frequency data are to be substituted as a test of the model. The states are not a priori defined, they could be, for example, nine men in a prison planning to escape, or nine trials by one person doing a Sudoku puzzle.

(*b*) Bottom-up activity involves sensory-motor extrinsic awareness of the environment leading on to Top-up arousal awareness of the self. The first explicit reference to Top-down Bottom-up appears to be by Siegel, Körding & König (2000) where somato-dendritic interactions are focused. An extension to Bayesian modeling in this context was made by Vilares and Kording (2011). The modeling of top-down and bottom-up has now become widely explored in theoretical psychology, for example by Shea (2013). The workspace is sometimes called CNN (chaotic neural networks) which can imply fractal dynamics arising when top-down dynamics are active.

The role of top-down bottom-up models was emphasized by Del Cul, Dehaene, Reyes Bravo & Slachevsky (2009) where on page 2537 they wrote "the exchange of top-down and bottom-up signals between frontal networks and posterior visual areas plays an essential role in access to consciousness" where both are continuously involved in feed-back loops based on neural networks and different PFC (pre-frontal cortex) brain regions. See also Ciszak et al (2013).

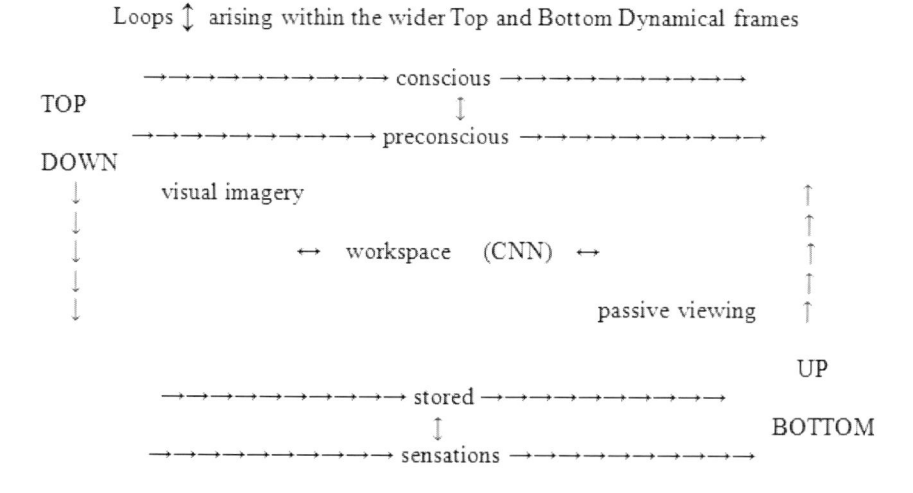

Figure 2. Qualitative processes represented in diverse neurophysiological regions.

Mechelli, Price, Noppeney, & Friston (2003, p.930) added another possible relation between top-down and bottom-up mediation and the use of category effects. Category effects can be mediated either by top-down or bottom-up mechanisms depending on the context. Passive viewing was mediated by bottom-up whereas visual imagery used top-down respectively. (see Figure 2 diagram above) Category effects in the early visual cortex leading into the occipital and temporal context can be mediated by bottom-up processes.

Julia Sets Derived from Equations 1 and 2

(*c*) We have here two Julia sets, a typical print of the Mandelbrot figure in two dimensions (a rich variety of software for exploring and printing these figures, in various scales and colours, is freely available in Wikipedia, and one local area from Eq. 2., in Figure 4.

Vrobel (2011, p. 129) shows an example, created by Rodd Halstead using a quartic equation. The spirals are numerous in this Figure 3 from a Wikipedia sources version.

The main X - axis of the Julia set we have here set obliquely, the perfect symmetry is shown about the axis, and the fractal repeating small extensions in the vertical axes in some various scales may be explored in more detail if the scale of the figure is increased. It interestingly also displays synchronization on axes that are oblique, and not parallel to or orthogonally to the main axis.

Figure 3. The Mandelbrot form exists showing two regions existing outside the black hole, with synchronization over scales or with converging spirals running to extinction.

A part of the Julia set of Eq. 2, showing a region where symmetrical pairs with synchronization are locally near to two parts of spirals, those originating or ending on a single point. It is perhaps a puzzle to some readers as to where these spirals come from. In more precise terminology, they are logarithmic spirals, Vitiello (2012a) stated "the microscopic nature of fractals appears to emerge from microscopic coherent local deformation processes. The dissipative structure of fractals is related to noncommutative geometry." Vitiello (2012b) presents a detailed account of the geometric properties of the logarithmic spiral in polar coordinates, for both left- and right-hand forms, or what are called clockwise and anti-clockwise. The simplest way to generate a single logarithmic spiral is to use polar coordinates, when the generating equation with using (r, θ) is

$$r = r_0 e^{d_0} r_0 > 0 \tag{3}$$

Figure 4. Parts of symmetrical pairs of logarithmic spirals built from Eq. 2.

In the limit it goes to either a circle of radius zero, or to a straight half-line. We can measure how tight the spiral is. This Eq. (3) does not contribute to predicting the existence of spirals in matching pairs, either in a sort of parallel configuration in Figure 4 or diagonal matching as in Figure 3. Also the spirals are not single equations without apparent noise, but have apparently been loaded with richly structured fractal dynamics. One cannot simply derived the Eq. (3) polar form from either of the time series fundamentals in Eq. (1) and (2), but if those Eqs. are treated as the bases of bottom-up into CNN, then the top-down loads the generating picture of Julia sets, as in Mandelbrot examples cited by Vrobel (2011) and the Gamma Eq. (2).

Discontinuities in Various Forms of Transient Synchronous Series

Synchronization is not a prior process, but arises as a derivative from transient non-linear dynamics in two or more dimensions generated within time-series, as Δ_t depth.

There is not one sort of synchronization, but as listed by Qian (2014, p. 1) several types include "complete synchronization, weak synchronization, lag synchronization, phase synchronization, and generalized synchronization". Benderskaya.& Zhukova (2011) extended these types a bit more, as being fragmentary over time.

The discovery of synchronization transitions, with abrupt and irreversible phase transitions to coupled oscillators, have been used to model biological systems, such as fireflies, brain neuron networks, cardiac pacemaker cells, and electrical engineering applications. Synchronization transitions are induced by the presence of both time delay and long-range connections.

Qian (p.8) found four distinct parameter regions, that are: an asynchronous region, a transition region, a synchronization region, and an oscillating region, as time delay is increased. If we believe that the relations between some fMRI and cognitive processes such as memory and decision are sufficiently mapped, then fractal dynamics are involved in cognition in a wide sense.

The information in Eq. 1 and Eq. 2 and in Figure 1 and Figure 2 is about time series movements, but in Figure 3 there two other types of series, going to extinction. On the edges of the black hole regions are fractal synchronized replicated copies in progressive shrinking scales, and over an extended area are series in strongly converging spirals leading to a point at terminal extinction. An illustration of parts of such spiral convergences is shown on page viii in the book by Gregson (1995), as part of the Julia set outlying regions in Eq. 2.

In the last two years a large collection of works has focused on what is now called Explosive Synchronization, particularly in adaptive and multilayer networks (Zhang, Boccaletti, Guan& Liu 2015). This mathematical situation appears to be more relevant to conjectures on collective human behavior than simpler statements about synchronization. Li et al (2013) said " processes like synchronization in networks have also been found to be significantly affected by the topological heterogeneity of the underlying network,…, explosive synchronization is not present in all heterogeneous networks." "In the presence of a positive correlation between the degree and the natural frequency of a node, the difference between a node and its neighbours in terms of their degrees can dramatically influence the corresponding synchronization process... For explosive synchronization to take place, the level of heterogeneity of the network cannot be too low to ensure the existence of the difference in the intrinsic frequencies of the nodes". Vlasov, Zou & Pereira (2014, p.2) report two results: "transitions associated with explosive synchronization are discontinuous, and there is a hysteresis loop associated with explosive synchronization".

Another variant of explosive synchronization is related to the network topology and intrinsic dynamics of an oscillator (Ji et al, 2013) where there is a microscopic correlation between the topology and its dynamics. The mathematical model used is intrinsically much more complex than Eq. 1 and Eq. 2, requiring 10 equations in the derived Kuramoto (1984) model built by Ji et al. We cannot readily know what a transient and explosive situation is, if the data are only derived from records of a social group of human or animal behavior. We need to know exactly when some examples of the explosions happen.

Singularity in Time and in Processing

Singularity has zero derivatives, as:

$$\partial Y/\partial j \Rightarrow 0 \quad \partial_2 Y/\partial j^2 \Rightarrow 0 \tag{4}$$

It is possible that the repeated transient gaps (Pöppel, 1987) in consciousness time series can be singularities, as remarked in (C, NC) in Fig 1.

Three possible dynamics are as here listed, partly described by Gregson (2013):

- Transitions across singularities, arising in macroscopic quantum systems.
- Singularities across bifurcations, with parameters not crossing over the bifurcations.
- Singularities in pairs of reflected synchronized transitions each made unidirectionally; that can be found, for example, in cusp catastrophies.

The problem we face is to see how we progress, via bottom up dynamics, from the time serious stimuli algebra at Eqs. 1 and 2, via CNN, to NC level and then down bottom to become the pictures in the form of multiple logarithmic spirals with fractal accretions, without arising in phenomenological consciousness (i.e., awareness).

If local very small discontinuities are written into series such as Eqs. 1 or 2, either randomly or with some periodicity and frequency , then we do not know *a priori* what will result in Figs 3 or 4, so there is much still to be done if the mathematics is to be accepted as a credible description of some human brain network dynamics.

CONCLUSION

There is a curious dynamical parallel between some contrasts in cosmology, where Big Bang in one universe starting from, or shrinking eventually to nothing, has been challenged, for example by Tegmark (2014) or Stenger (2015), by the idea of a Multiverse with parallel continuity of many universes, mutually undetectable, and the scene in Fig 4 where we have both parallel synchronized states, one in the lower of our image, and also the spiral patterns at the upper right expanding from or contracting to a point, which might be thought to echo the Big Bang. Vitiello (2012a,b) also notes that the geometry of a logistic spiral has analogues in cosmology. This is just a friendly speculation for the reader to enjoy, to remind us of analogous unsettled questions both in cosmology and in consciousness.

REFERENCES

Becker, K.-H.& Dörfler, M. (1988) *Dynamische Systeme und Fraktale* [Dynamical Systems and Fractals]. Braunschweig: Vieweg.

Beggs, J. M.& Plenz, D. (2003) Neuronal Avalanches in Neurocortical Circuits. *The Journal of Neuroscience,* 23(35), 11167- 11177.

Benderskaya, E. N. & Zhukova, S. V. (2011) Chaotic Clustering: Fragmentary Synchronization of Fractal Waves. In: Esteban, T.-C. (Ed.) *Chaotic Systems.* In Tech 2011, pp. 187-202.

Ciszak, M., Euzzor, S., Geltrude, A., Arecchi, F. T. & Meucci, R. (2013) Noise and coupling induced synchronization in a network of chaotic neurons. *Communications in Nonlinear Science and Numerical Simulation,* 18, 938-945.

Crick, F. & Koch, C. (2003) A framework for consciousness. *Nature Neuroscience*, 6, (2), 119-126.

Del Cul, A., Dehaene, S., Reyes, P., Bravo, E. & Slachevsky, A. (2009) Causal role of prefrontal cortex in the threshold for access to consciousness. *Brain,* 132, 2531 – 2540.

Engel, A. K., Fries, P., König, P. Brecht, M. & Singer W. (1999) Temporal Binding, Binocular Rivalry, and Consciousness. *Consciousness and Cognition,* 8, 128-151.

Gregson, R. A. M. (1992) *n-Dimensional Nonlinear Psychophysics.* Hillsdale, NJ.: Lawrence Erlbaum Associates.

Gregson, R. A. M. (1995) *Cascades and Fields in Perceptual Psychophysics.* Singapore: World Scientific.

Gregson, R. A. M. (2006) *Informative Psychometric Filters.*ANU e-mail.

Gregson, R. A. M. (2013) Symmetry-Breaking, Grouped Images and Multistability with Transient Unconsciousness. *Nonlinear Dynamics, Psychology and Life Sciences,* 17 (3), 325-344.

Ji, P., Peron, T. K., Menck, P. J., Rodrigues, F.A. & Kurths, J. (2013) *Cluster Explosive Synchronization in Complex Networks.* arXiv:1303.3498v2 [nlin.AO]

Kozma, R. & Freeman, W. J. (2008) Intermittent spatio-temporal desynchronization and sequenced synchrony in ECoG signals. *Chaos, An interdiscioplinary Journal of Nonlinear Science,* 18, 037131, doi.org/10.1063/1.2979694.

Kuramoto, Y. (984) *Chemical Oscillations, Waves and Turbulence.* New York, N.Y: Springer Verlag.

Li, P., Zhang, K., Xu, X., Zhang, J., & Small, M. (2013) Reexamination of explosive synchronization in scale-free networks: The effect of disassortativity. *Physical Review E,* 87, 042803, 1-5.

Mandelbrot, B. B. (2004) *Fractals and Chaos: The Mandelbrot Set and Beyond.* New York: Springer Verlag.

Mechelli, A., Price, C. J., Friston, K. J. & Ishai, A. (2004) Where Bottom-up Meets Top-down: Neuronal Interactions during Perception and Imagery. *Cerebral Cortex,* 14, 1256-1265.

Nicolis, G. & Basios, V. (Eds.) (2015) *Chaos, Information Processing and Paradoxical Games.* Singapore: World Scientific.

Nieuwenhuis, I. (2004) Synchronization of brain activity in access consciousness. *Psychonomics and Neurobiology,* Literature Study, 1-17. University of Amsterdam.

Pöppel, E. (1987) *Grenzen des Bewußtseins* [Limits of consciousness]. München: dtv Sachbuch.

Qian, Y. (2014) Time Delay and Long-Range Connection Induced Synchronization Transitions in Newman-Watts Small-World Neuronal Networks. *PLoS ONE* 9(5): *e96415:* doi: 10.1371/journal.pone.0096415

Seth, A.K., Suzuki, K. & Critchley, H.D. (2012) An interoceptive predictive coding model of conscious presence. *Frontiers in Consciousness Research,* 2, 395-424.

Shea, N. (2013) Distinguishing Top-Down from Bottom-Up Effects. In: Biggs, S., Matthen, M. & Stokes, D. (eds). *Perception and its Modalities,* Oxford: OUP.

Siegel, M., Körding, K.P. & König, P. (2000) Integrating top-down and bottom-up sensory processing by somato-dendritic interactions. *Journal of Computing and Neuroscience,* 8(2), 161-173.

Singer, W. & Gray, C. M. (1995) Visual feature integration and the temporal correlation hypothesis. *Annual Review of Neuroscience,* 18, 555-586.

Singer, W. (2001) Consciousness and the binding problem. *Annals of the New York Academy of the Sciences,* 929, 123-146.

Stenger, V. (2015) *God and the Multiverse.* New York: Prometheus Books.

Tegmark, M. (2014) *Our Mathematical Universe.* New York: Alfred A. Knopf

Treisman, A. (1998) Feature binding, attention and object perception. *Philosophical Transactions of the Royal Society of London,* B. 353, 1295-1306.

Tufte, E. R. (2001/2011) *The Visual Display of Quantitative Information. (2nd. Edn.)* Cheshire, Conn: Graphics Press.

Vitiello, G. (2012a) Fractals, coherent states and self-similarity induced noncommutative geometry. *Physics Letters A,* 376, 2527-2532.

Vitiello, G, (2012b) Fractals, Dissipation and Coherent States. *Quantum Interaction, Lecture Notes in Computer Science,* 7620, 68-79.

Vilares, I. & Kording, K. (2011) Bayesian models: the structure of the world, uncertainty, behavior, and the brain. *Annals of the New York Academy of the Sciences,* 1224, 22-39.

Vlasov, V, Zou, Y. & Pereira, T. (2014) Explosive Synchronization is Discontinuous. *Nonlinear Sciences: Adaptation and Self-organizing Systems.* arXiv:1411.6873v1, [nlin.AO]

Vrobel, S. (2011) *Fractal Time: Why a Watched Kettle Never Boils.* New Jersey: World Scientific.

Zakharova, A., Kapeller, M. & Schöll, E. (2014) Chimera Death: Symmetry Breaking in Dynamical Networks. *Physics Review Letters,* 112 (15), 154101.

Zhang, X., Boccaletti, S., Guan, S. & Liu, Z. (2015) Explosive synchronization in adaptive and multilayer networks. *Physics Review Letters,* 114 (3), 038701.

In: Chaos and Complexity in the Arts … ISBN: 978-1-53612-995-3
Editors: N. Sala and G. Cappellato © 2018 Nova Science Publishers, Inc.

Chapter 12

NEW KOCH CURVES AND THEIR CLASSIFICATION

Mamta Rani[*1], *Riaz Ul Haq*[2] *and Norrozila Sulaiman*[2]
[1]Krishna Engineering College, Ghaziabad, India
[2]Faculty of Computer Systems and Software Engineering,
University Malaysia Pahang, Gambang, Kuantan, Malaysia

ABSTRACT

The von Koch curve, a classical fractal model, is generated by dividing the initiator into three equal parts. There exist a few numbers of variants of Koch curve in literature. In this paper, we give new Koch curves generated by dividing the initiator into unequal parts. With the increase in size of the set of Koch curve, we felt that there is a need of classification of Koch family. The classification is based on their method of generation. Further, an application of new Koch curves in wireless communication has been proposed.

Keywords: Koch curve, Koch snowflake curve, superior iterates, superior Koch curve, classification, fractal dimension

---

[*] E-mail: mamtarsingh@rediffmail.com

1. INTRODUCTION

The Koch snowflake (also known as the Koch star and Koch island) is a mathematical curve, which is continuous everywhere but differentiable nowhere. It is an example of bounded curve of infinite length [20, p.13], [6, p. xxiii]. Koch snowflake curve is used as a fractal antenna. Cohen [3] described the importance of fractal antennas, explicitly, for wireless technologies, especially in military services. For a detailed study on Koch fractal antenna and various researches on it, one may refer to [4, 7, 9, 10, 12, 21, 22, 24, 25] and sever cross references thereof. McClure [11] has discussed the vibrational modes of a drum, shaped like Koch snowflake. Epstien and Adeeb [5] derived stiffness of the Koch curve. Barcellos [1] gave variants of Koch curve by dividing the initiator into 4 equal parts. Vinoy et. al. [23] also presented a new way of generation of variants of Koch curve by varying indentation angle and gave a formula to calculate their fractal dimension.

In 2004, Rani and Kumar [16, 17] implemented two-step feedback machine via superior iterates and computed superior fractals. To see the applications of superior iterates in fractal theory, one may refer to Rani and Agarwal [14] and its further details of usefulness in cross references thereof. The concept of two-step feedback machine also gave idea of generation of fractal objects by unequal division of initiator. Using this idea, Rani et. al. computed superior fractal carpets [15], superior fractal plants [2] and superior Cantor sets and superior Devil's staircases [18], and presented their categorizations.

In the series of superior fractals, authors are working in superior Koch curves. Authors have computed two new shapes of Koch curve by dividing initiator into the line segments of unequal length [8]. In this paper, we have generated new Koch curves and presented their classification. In Section 2, basic construction of Koch curve and definition of superior iterates have been described. In Section 3, categorization of Koch curves have been presented. Further, an application of new Koch curves in antenna theory has been proposed in Section 4, followed by concluding remark in Section 5.

2. PRELIMINARIES

To generated the von Koch curve, a line segment is divided into three segments r_1, r_2 and r_3 such that $r_1 = r_2 = r_3$, and the middle line segment is manipulated as shown in Figure 1a. The same method of division and manipulation is applied on the four small line segments, obtained after first iteration, infinitely. The fractal dimension of the von Koch curve is ≈ 1.262. When Koch curve is generated on initiator triangle, we obtain Koch snowflake curve (see Figure 1b) [6], [13], [20], which is used as a fractal antenna.

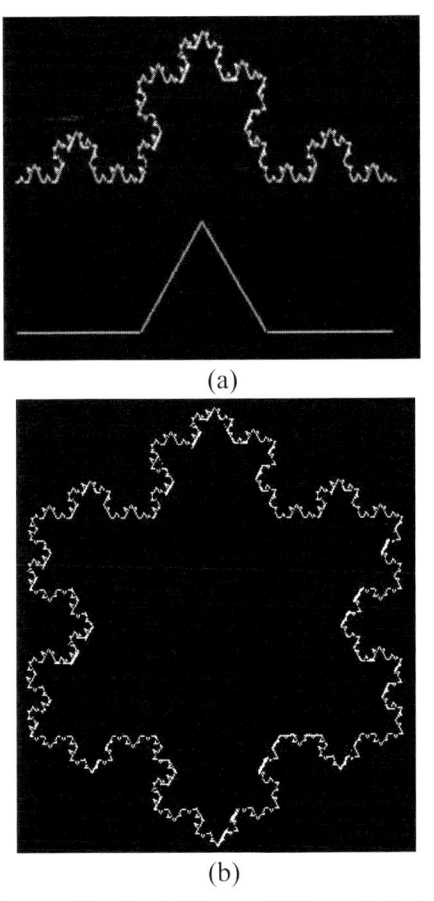

(a)

(b)

Figure 1. (a) von Koch curve (Koch middle one-third curve) for $(r_1, r_2, r_3) = (1/3, 1/3, 1/3)$ with its generator. (b) Koch snowflake antenna (Koch middle one-third snowflake curve).

Superior iterates: Let A be a subset of real or complex numbers and F: $A \rightarrow A$. For $X_0 \in A$, a sequence $\{X_N\}$ in A is construct in the following manner:

$$X_N = B_N F(X_N - 1) + (1 - B_N) X_{N-1},$$

where $0 < B_N \leq 1$ and $\{B_N\}$ is convergent away from zero. The sequence $\{X_N\}$ is known as superior iterates. Superior iterates are an example of two-step feedback machine. At $B_N = 1$, it behaves as one-step feedback machine [16, 17]. Rani et. al. [2, 15, 18] obtained the idea of dividing initiator into unequal parts from superior iterates and generated superior fractals.

3. CLASSIFICATION OF NEW KOCH CURVES

Inspired by superior iterates, we compute the new Koch curves. These curves are new examples of superior fractals. Now, we present different variants of Koch curve in two categories. The classification is based on the method of generation of curves.

Category 1: Generation by Equal Division

In this category, we divide the initiator into three equal parts and manipulate one of the segment. Repeat the same process on the remaining segments sufficient number of times. Conventional Koch curve is an example of this category. Koch left one-third curve (Figure 2a) is another new example of Category 1, which is generated by dividing the initiator into three equal parts and manipulating the left segment. Fractal dimension of this curve is $\approx$ 1.262. Also, see Koch left one-third snowflake curve, when initiator is a triangle.

Similarly, Koch right one-third curve may also be computed as another example of Category 1. Following the same method of nomenclature, authors have renamed the conventional Koch curve as Koch middle one-third curve [8], which again falls in Category 1.

There are many other variants of Koch curve available in literatures that are fit into this Category. Barcellos [1] generated variants of Koch curve of dimension between 1 and 2 by dividing the initiator into four equal parts. He manipulated middle two segments such that one can obtain 3, 4, 5 or 6 components; each of them is of length ¼ of the initiator. As the number of components of equal length increases in Barcello's Koch curves, their dimension increase.

As per our nomenclature method, we give a name to Barcello's Koch curves as Koch middle two one-fourth curves. Vinoy et. al. [23], also, modified the Koch curve by altering its indentation angle.

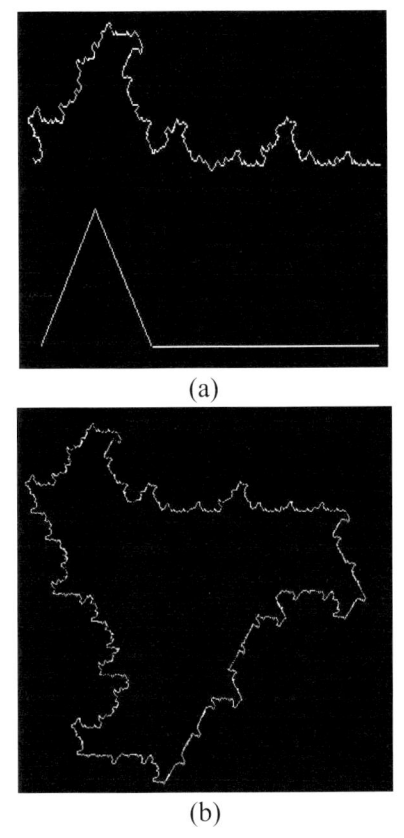

(a)

(b)

Figure 2. (a) Koch left one-third curve for $(r_1, r_2, r_3) = (1/3, 1/3, 1/3)$ with its generator. (b) Koch left one-third snowflake curve.

Category 2: Generation by Unequal Division

Now, we consider Koch curves obtained by unequal division of the initiators. We present two cases in this category.

Case 1: In this case, we divide the initiator into three parts such that two parts are equal in length and 3rd part is unequal, and manipulate one of the segments iteratively. We present the Koch right half curve as an example of Case 1.

To draw the same (Figure 3a), divide the initiator line into three parts such that left and middle line segments are one-fourth and right line segment is half of the initiator respectively. Now, manipulate the right line segment as shown in Figure 3a.

Repeat the same process on the remaining four segments sufficient number of times. Fractal dimension of the Koch right half curve is ≈ 1.2925, which is greater than the Koch curves of Category 1.

See also, Figure 3b for Koch right half snowflake curve generated on initiator triangle. Koch middle half curve [8], middle one-forth and middle one-eighth curves [19] also fall in Case 1.

Case 2: In this case, we divide the initiator into three parts such that the all parts are unequal in length. Koch middle one-sixth curve (Figure 4a) is an example of Case 2. To draw the same, divide the initiator line into three parts such that left, middle and right segments are 1/3, 1/6 and ½ of the initiator respectively. Now, manipulate the middle line segment as shown in Figure 4a. Repeat the same process on the remaining four line segments sufficient number of times. Fractal dimension of the Koch middle one-sixth curve is ≈ 1.086, which is less than the Koch curves of Category 1.

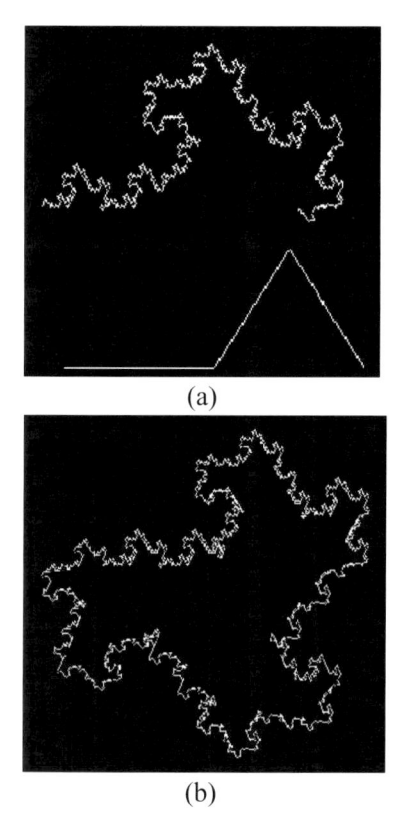

(a)

(b)

Figure 3. (a): Koch right half curve for $(r_1, r_2, r_3) = (1/4, 1/4, 1/2)$ with its generator. (b): Koch right half snowflake curve.

We obtain Koch middle one-sixth snowflake curve (Figure 4b) by applying the above process on a triangle. The Koch middle one-fifth curve [8] is another example of Case 2.

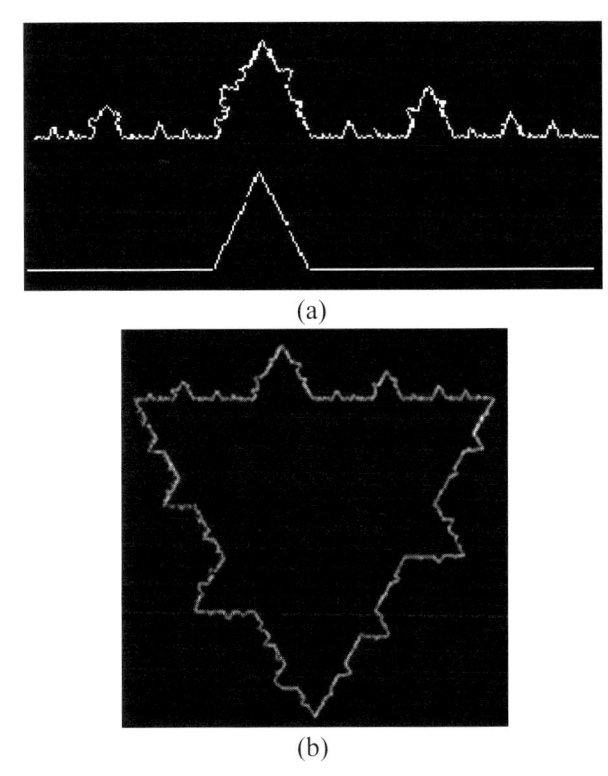

(a)

(b)

Figure 4. (a): Koch middle one-sixth curve for $(r_1, r_2, r_3) = (1/3, 1/6, 1/2)$ with its generator. (b): Koch middle one-sixth snowflake curve.

4. APPLICATION OF SUPERIOR KOCH CURVES

The new Koch curves can be used as fractal antenna in wireless technology. According to Vinoy, Jose and Varadan [22], since fractal antennas having smaller dimensions show better multi-band characteristics. Therefore, the Koch middle one-fifth and Koch middle one-sixth curves of Category 2 that have lesser dimension than conventional Koch curve will, possibly, show better multi-band characteristics as fractal antennas.

Vinoy, Jose and Varadan [23] generated new shapes of Koch curve by varying its indentation angle, while we have generated new Koch curve by unequal division of the initiator. This difference raises the following curiosity: Whether multi-band characteristics of the two antennas, having same dimension but different method of generation, will be different?

CONCLUSION

Using the idea of division of initiator into unequal parts [2, 15, 18], a few variants of Koch curve have been presented in this paper. These Koch curves are new examples of superior fractals. We have classified different variants of Koch curve into two categories. Category 1 includes the Koch curves that are generated by dividing the initiator into equal parts and Category 2 includes Koch curves that are generated by the division of initiator into unequal parts. In both the categories, following the given ideas, many Koch curves can be generated. In this paper, we surveyed for other different types of variants of Koch curve also, and made them fit into our classification.

REFERENCES

[1]   Barcellos, Anthony, The Fractal geometry of Mandelbrot, *The College Mathematics Journal,* 15(2), 1984, 98-114.

[2]   Chandra, M. and Rani, M., Categorization of fractal plants, *Chaos, Solitons, Fractals*, 41(3), 2009, 1442-1447.

[3]   Cohen, N., *Fractals' new era in military antenna*, 1996. http://mobiledevdesign.com/hardware_news/radio_fractals_new_era/

[4]   Elkamchouchi, Hassan, Nasr, And Mustafa Abu, 3d Fractal Rectangular Koch Dipole And Hilbert Dipole Antennas, *Ieee Int. Conf. On Microwave and Millimeter Wave Technology*, 2007, 1 - 4.

[5]   Epstien, M., and Adeeb, Sameer M., The stiffness of self-similar fractals, *Int. J. Solids Struct.*, 45, 2008, 3238-3254.

[6]   Falconer, K., *Fractal Geometry: Mathematical foundations and applications* (second edition), John Wiley and Sons, Hoboken, NJ, 2003. MR2118797

[7]   Ghatak, Rowdra, Poddar, Dipak R., And Mishra Rabindra K., A Moment-Method Characterization Of V-Koch Fractal Dipole Antennas, *Int. J. Electron. Commun.* (AEU), 63, 2009, 279-286.

[8]   Haq, Riaz Ul, Sulaiman, N., and Rani, M., Superior fractal antennas, *Malaysian Technical Universities Conference on Engineering and Technology*, Melaka, Malaysia, Jun 28-29, 2010, 23-26.

[9]   Kordzadeh, A., And Kashani, F. Hojat, A New Reduced Size Microstrip Patch Antenna With Fractal Shaped Defects, *Progress In Electromagnetics Research B*, 11, 2009, 29-37.

[10] Krishna, D. D., Gopikrishna, M., Anandan, C. K., Mohanan, P., and Vasudevan, K., Compact Wideband Koch Fractal Printed Slot Antenna, Iet Microw. *Antennas Propag.*, 3(5), 2009, 782-789.

[11] McClure, M., Images of a vibrating Koch drum, *Computers and Graphics*, 32, 2008, 711-715.

[12] Mirzapour, B., and Hassani, H. R., Size reduction and bandwidth enhancement of snowflake fractal antenna, IET Microw. *Antennas Propag.*, 2(2), 2009, 180-187.

[13] Peitgen, H. O., Jürgens, H. and Saupe, D., *Chaos and Fractals: New frontiers of science* (second edition), 2004, Springer-Verlag, New York. MR2031217 Zbl 1036.37001.

[14] Rani, M., and Agarwal, R., Effect of stochastic noise on superior Julia sets, *J. Math. Imaging and Vis.*, 36, 2010, 63-68.

[15] Rani, M., and Goel, S., Categorization of new fractal carpets. *Chaos, Solitons, Fractals*, 41(2), 2009, 1020-1026.

[16] Rani, M., and Kumar, V., Superior Julia set, *J. Korea Soc. Math. Educ., Ser. D; Res. Math. Educ.*, 8(4), 2004, 261-277.

[17] Rani, M., and Kumar, V., Superior Mandelbrot set, *J. Korea Soc. Math. Educ., Ser. D; Res. Math. Educ.*, 2004, 8(4), 279-291.

[18] Rani, M., and Prasad, S., Superior Cantor sets and superior Devil's staircases. *Int. J. Artif. Life Res.*, 1(1), 2010, 78-84.

[19] Sengufta, Kaushik, Influence of fractal lacunarity on the performance of dipole antennas with generalized Koch curves, Int. Conf. Applied Electromagnetic sand Communications, *IECCom* 2005, Oct 12-14, 1- 4.

[20] Schroeder, M., *Fractals, Chaos, Power Laws: Minutes from an infinite paradise*, 1991, W. H. Freeman and Company, New York. MR1098021.

[21] Song, X. D., Fu, J. M., and Wang, W., Design of a miniaturized dual band Koch fractal boundary microstrip antenna, *IEEE Microwave Conference*, 2008, 282 - 284.

[22] Vinoy, K. J., Jose, K. A., and Varadan, V. K., Impact of fractal dimension in the design of multi-resonant fractal antennas, *Fractals*, 12, 2004, 55-66.

[23] Vinoy, K. J., Jose, K. A., and Varadan, Multi-band characteristics and fractal dimension of dipole antennas with Koch curve geometry, *Antennas and Propagation Society International Symposium,* 4, 2002, 106-109.

[24] Werner, Douglas H., and Ganguly, Suman, An overview of fractal antenna engineering research, *IEEE Antennas and Propagation Magazine,* 45(1), 2003, 38-57.

[25] Zhang, Yizhe, and Kishk, Ahmed A., Analysis of dually polarized fractal antennas, *Antennas and Propagation Society International Symposium,* 2006, 2041ss - 2044.

In: Chaos and Complexity in the Arts …   ISBN: 978-1-53612-995-3
Editors: N. Sala and G. Cappellato   © 2018 Nova Science Publishers, Inc.

Chapter 13

NEW SIERPINSKI CURVE JULIA SETS

Manish Kumar[1,*] and Mamta Rani[2,†]

[1]DNM Institute of Engineering & Technology, Lucknow, India
[2]Central University of Rajasthan, Ajmer, India

ABSTRACT

Rational maps and its generalizations have been evolved extensively using the function iteration. In this paper, we introduce superior iterations in the study of Julia sets for rational maps of the form $f\lambda(z) = z^2 + \lambda / z^2$ when $\lambda \neq 0$. Our new Sierpinski curve Julia sets are effectively different from those obtained by others.

Keywords: rational maps, Sierpinski Julia sets, superior iterations, superior Sierpinski Julia sets

1. INTRODUCTION

In this paper, we discuss the one-parameter family of rational maps

$$f_\lambda(z) = z^2 + \lambda / z^2$$

---

[*] E-mail address: dr.manish.2000@gmail.com
[†] E-mail address: mamatarsingh@gmail.com (Corresponding author)

where $\lambda \neq 0$ is a complex parameter.

The behavior of the orbit arising out of the function iteration of $f_\lambda(z)$ for $\lambda = 0$, is simple and well understood. In this case, Julia set is the unit circle. If $0 < |z| < 1$, then $f_\lambda(z) < |z|$, and all iterations that begin inside the unit circle simply tend to 0, which is an attracting fixed point. If $|z| > 1$, then iterates increase in magnitude and tend to ∞. When $\lambda \neq 0$, then the function $f_\lambda(z)$ is a rational map of degree four. In this case, we see the dramatic and interesting changes in the dynamic of $f_\lambda(z)$ (see, for instance, Blanchard [1], Blanchard *et al.* [2], Devaney [3], Devaney *et al.*, [5]-[9], Milnor [12] and Milnor and Tan Lei [13]).

Devaney *et al.* ([3], [4], [7] and [9]) have investigated various aspects of Julia sets for the rational maps when the parameter λ is nonzero but small. In 2005, Blanchard *et al.* [2] considered the family of rational maps $f_\lambda(z) = z^2 + \lambda / z^2$, and proved that in any neighborhood of $\lambda = 0$ in the parameter plane, there are infinitely many disjoint open sets of parameters for which the Julia set is a Sierpinski curve [3]. Devaney [4] also produced some fascinating and marvelous pictures of Julia sets for different values of λ. Devaney and Look [7] have discussed the dynamics of $f_\lambda(z)$ when $|\lambda|$ is close to 0. Indeed, their main goal is to present a criterion for the Julia set of $f_\lambda(z)$ to be Sierpinski curve (see [7, p.2]).

In 2007, for the same family of rational maps, Devaney *et al.* [9] have considered the special case where the critical orbits for these maps are all eventually periodic. (A point x is an eventually periodic point of the function f with point k, If there exists an N such that $f^{n+k}(x) = f^n(x)$, whenever $n \geq N$ (see [14]). Such maps are often called Misiurewicz rational maps [9]. These maps showed that the Julia set of any such map is a generalized Sierpinski gasket [9], when these critical orbits also lie on the boundary of the basin of ∞.

Devaney *et al.* [5] described the dynamics of certain Julia sets of functions drawn from the family of rational maps of the complex plane given by

$$g_\lambda(z) = z^n + \lambda / z^n$$

where n and d are integers. When the parameters λ is nonzero and small, the map g_λ is called a *singular perturbation* of z^n. Several things happen when $\lambda \neq 0$. First of all, the map has degree $n + d$ rather than n. Second, the origin is a pole rather than a fixed point, and finally, there are $n + d$ new critical points in addition to the original critical points at 0 and ∞. (The 0 is the only pole for each function in the above family of rational maps).

In the generation of Julia sets, the escape criterion followed for all families of rational maps $f_\lambda(z) = z^2 + \lambda/z^2$ is $|z| > 2$ when $|\lambda| < 1$ with respect to the function iteration. (In a personal communication, the authors received this escape criterion from Prof. Robert L. Devaney, and they owe him for the same.)

In this paper, we introduce the superior iteration [14] in the study of Julia sets constructed by rational maps of the form $f_\lambda(z) = z^2 + \lambda/z^2$, where $\lambda \neq 0$. We have generated superior Sierpinski curve Julia sets and compared the same with those studied and obtained by Blanchard *et al.* [2] and Devaney *et al.* [9].

2. PRELIMINARIES

In all that follows

$f_\lambda(z) = z^2 + \lambda/z^2$ where $\lambda \neq 0$.

Filled Julia sets consists of those points which do not escape to infinity under the function iteration of $f_\lambda(z)$. It is well known the boundary of the filled Julia set is Julia set (cf. [3], [10], [12]).

By definition, a Sierpinski curve is a compact, connected, locally connected and nowhere dense subset of the plane that has the property that any two boundaries of complementary domains are pair wise disjoint simple closed curves (see [3, p.1]). The Sierpinski carpet is one of the well known, examples of Sierpinski curve ([3], [4], [13]).

Sierpinski curves arise as Julia sets of certain rational functions. The first example of the same was given by Milnor and Tan lei [13]. A more accessible collection of such Julia sets may be found in the family of rational functions given by $f_\lambda(z)$ for $\lambda \neq 0$ (cf. [4], [5], [6], [9]).

Definition 1 [2, 4]. Suppose that the critical orbit of f_λ tends to ∞ but the critical points of f_λ do not lie in the immediate basin of ∞. Then the Julia set of f_λ is a Sierpinski curve. Such Julia sets are known as Sierpinski curve Julia sets [2] and denoted by $J(f_\lambda)$.

In the family of rational maps $f_\lambda(z)$, there are infinitely open sets O_n on the negative λ-axis with $n \geq 2$ for which the following properties hold for each $\lambda \in O_n$:

(i) $J(f_\lambda)$ is a Sierpinski curve.

(ii) There is a unique attracting cycle for f_λ, namely the attracting fixed point at ∞.

(iii) The complementary domains in the Sierpinski curve Julia set are the components of the basin of attraction of ∞.

(iv) All four nonzero critical points of f_λ enter the immediate basin of attraction of ∞ at iteration n.

For details, one may refer to Devaney*et al.* [9].
We shall study the following iterated family of rational equation:

$z_{n+1} = \beta_n f_\lambda(z_n) + (1 - \beta_n) z_n, n = 1, 2, ...,$

where $0 < \beta_n \le 1$ and $\{\beta_n\}$ is convergent to β away from 0.

Definition 2. The sequence $\{z_n\}$ constructed above is called the superior sequence of iterates for $f_\lambda(z)$ with the initial choice z_0 and $\beta_n = \beta$, $n = 1, 2, ... 5$.

Notice that this definition yields the function iteration when $\beta_n = 1$, $n = 1$, 2, ... For a nice discussion of the above definition in the study of discrete dynamics, one may refer to [11], [14] and [15].

3. SUPERIOR SIERPINSKI CURVE JULIA SETS

The collection of points whose orbits are bounded under the superior iterations for the function $f_\lambda(z)$ is called the filled superior Sierpinski curve Julia set. The boundary of the filled superior Sierpinski curve Julia set is the superior Sierpinski curve Julia set (briefly denoted by $SJ(f_\lambda)$).

Following Rani [15], the superior escape criterion for the rational map $f_\lambda(z)$, we take

$\max\{|\lambda|, (2/\beta)\},$

where $0 < \beta \le 1$.

We have written a program in C++ to generate the Sierpinski curve Julia sets and superior Sierpinski curve Julia sets. We present some special figures for $f_\lambda(z)$ and compare them with their older brothers viz; Sierpinski curve Julia sets.

Fig 1(a). generated with the combination of $(\beta, \lambda) = (1, -0.0625)$. For the same value of λ with $\beta = 0.5$ and $\beta = 0.3$, Fig. 1(b) and Fig. 1(c) are generated respectively that are entirely different from Fig. 1(a) due to use of superior iterations in generations. A combination of $(\beta, \lambda) = (1, -0.01)$ gives Fig. 2(a). By combining the same λ with $\beta = 0.7$, Fig. 2(b) is produced. If we combine $\beta = 1$ (function iteration) and $\lambda = -0.25$, then we have Fig. 3(a). Fig. 3(b) is generated with the combination of $(\beta, \lambda) = (0.6, -0.25)$. Fig. 4(a) is produced by combining $\beta = 1$ and $\lambda = -0.001$. For the same value of λ with $\beta = 0.55$, Fig. 4(b) is generated. If we combine $\lambda = -0.01965 + i0.2754$ and $\beta = 1$, we have Fig. 5(a). By combining the same λ with $\beta = 0.75$, Fig 5(b) is produced. The combination of $(\beta, \lambda) = (1, -0.36428)$ generates Fig. 6(a), and finally Fig. 6(b) is generated for $(\beta, \lambda) = (0.87, -0.36428)$.

Figure 1(a). Sierpinski curve Julia set for $z^2 + \lambda/z^2$ with $(\beta, \lambda) = (1, -0.0625)$.

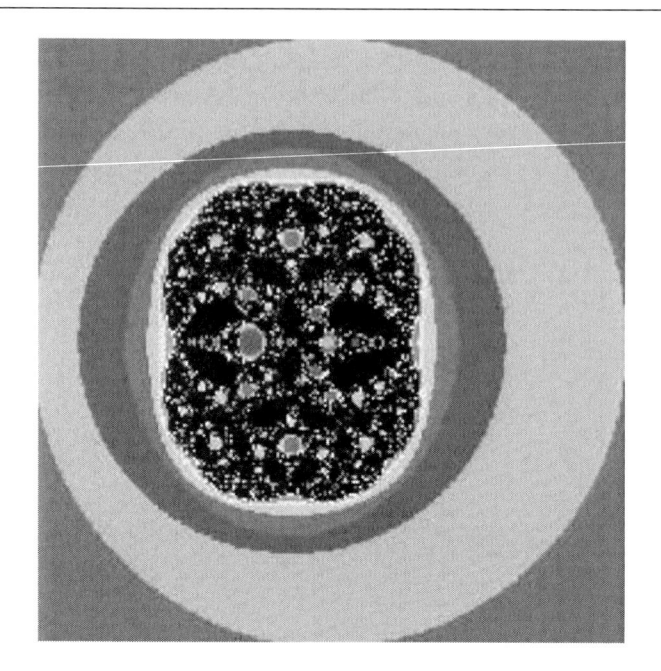

Figure 1(b). Superior Sierpinski curve Julia set for $z^2 + \lambda/z^2$ with $(\beta, \lambda) = (0.5, -0.0625)$.

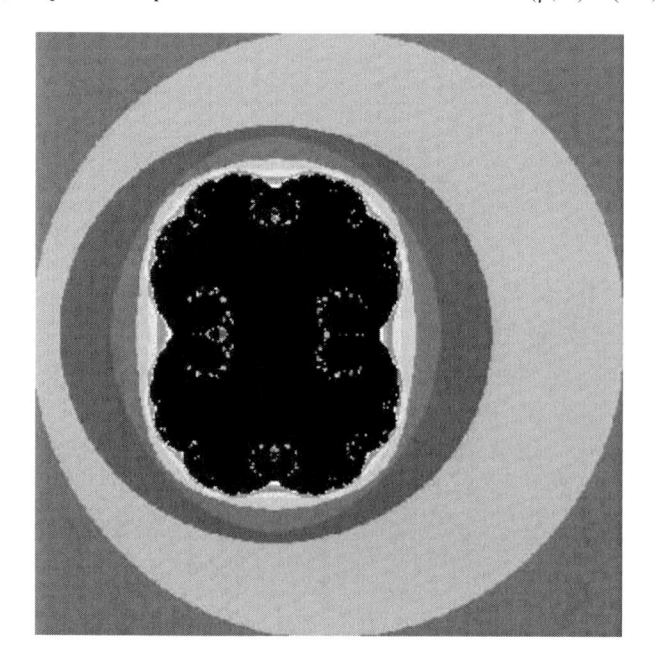

Figure 1(c). Superior Sierpinski curve Julia set for $z^2 + \lambda/z^2$ with $(\beta, \lambda) = (0.3, -0.0625)$.

Figure 2(a). Sierpinski curve Julia set for $z^2 + \lambda/z^2$ with $(\beta, \lambda) = (1, -0.01)$.

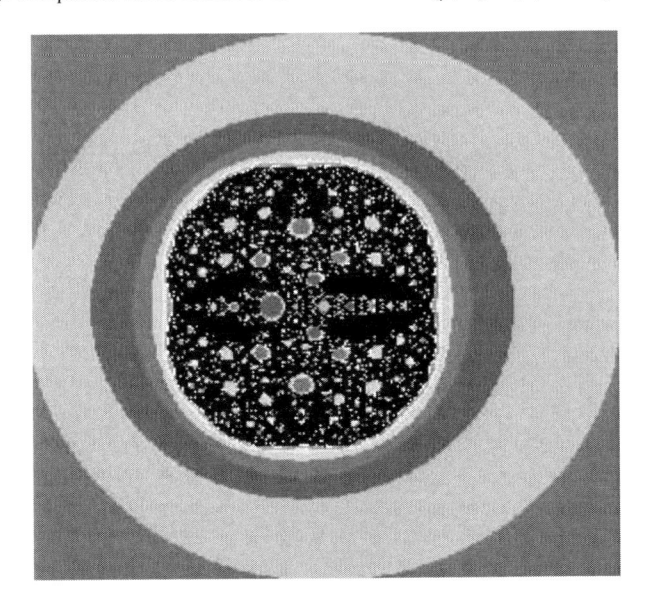

Figure 2(b). Superior Sierpinski curve Julia set for $z^2 + \lambda/z^2$ with $(\beta, \lambda) = (0.7, -0.01)$.

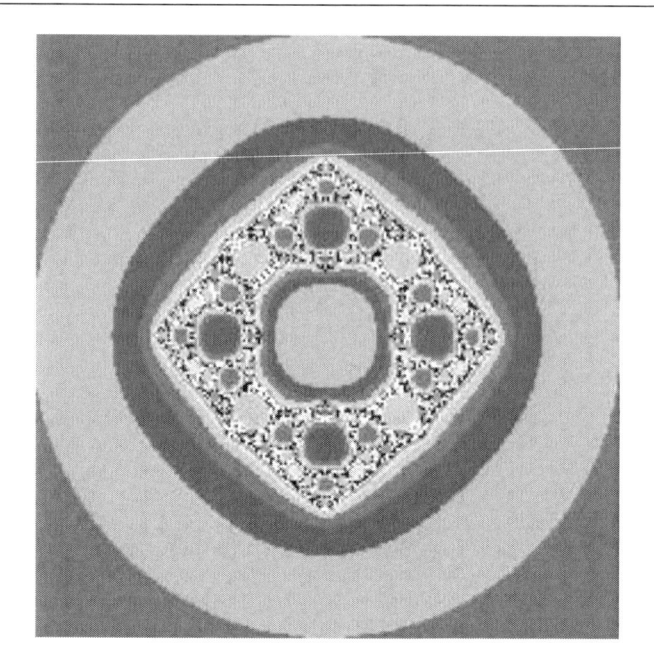

Figure 3(a). Sierpinski curve Julia set for $z^2 + \lambda/z^2$ with $(\beta, \lambda) = (1, -0.25)$.

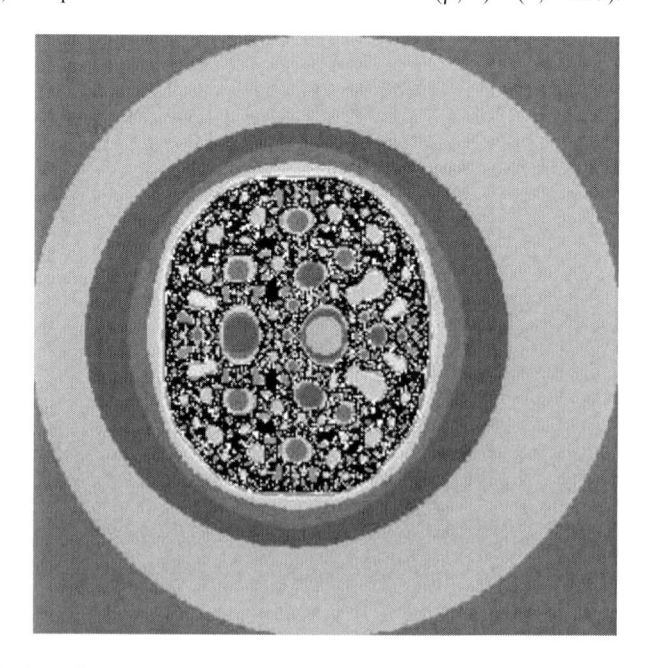

Figure 3(b). Superior Sierpinski curve Julia set for $z2 + \lambda/z2$ with $(\beta, \lambda) = (0.6, -0.25)$.

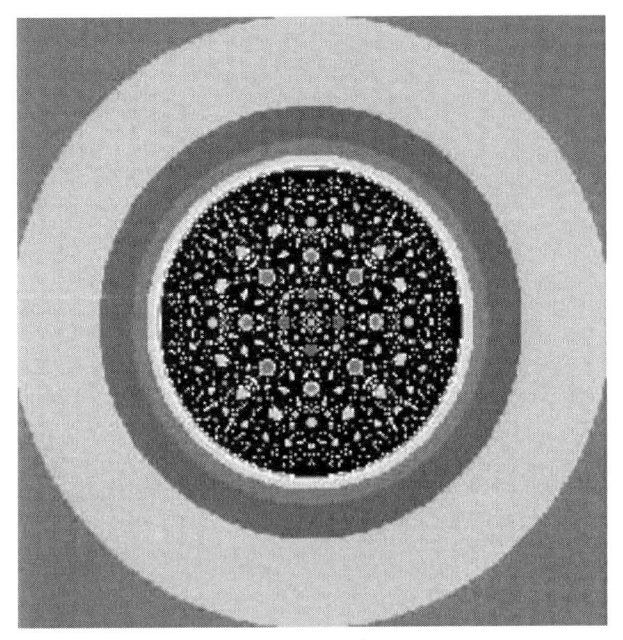

Figure 4(a). Sierpinski curve Julia set for $z^2 + \lambda/z^2$ with $(\beta, \lambda) = (1, -0.001)$.

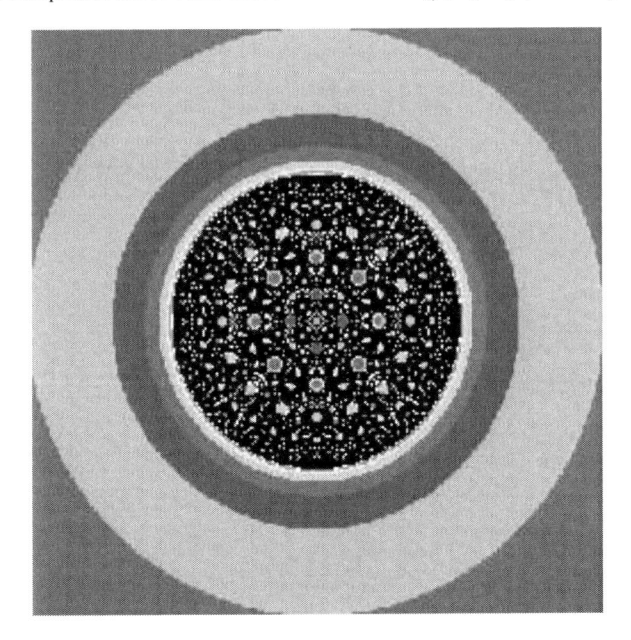

Figure 4(b). Superior Sierpinski curve Julia set for $z^2z^2 + \lambda/z^2$ with $(\beta, \lambda) = (0.55, -0.001)$.

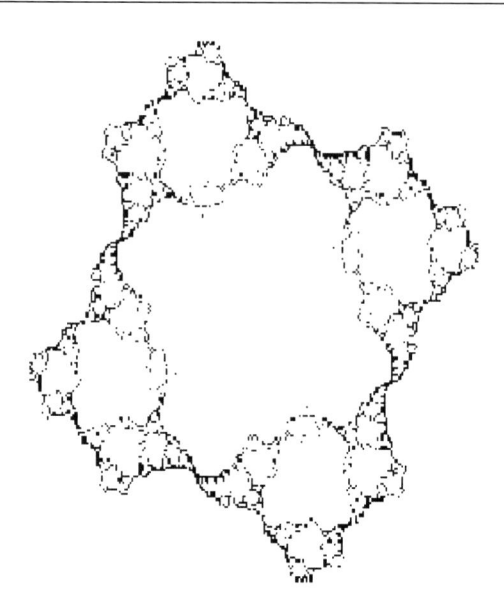

Figure 5(a). Sierpinski curve Julia set$z^2 + \lambda/z^2$with Sierpinski curve Julia set$z^2 + \lambda/z^2$ with $(\beta, \lambda) = (1, -0.01965 + i0.2754)$ $(\beta, \lambda) = (0.75, -0.01965 + i0.2754)$.

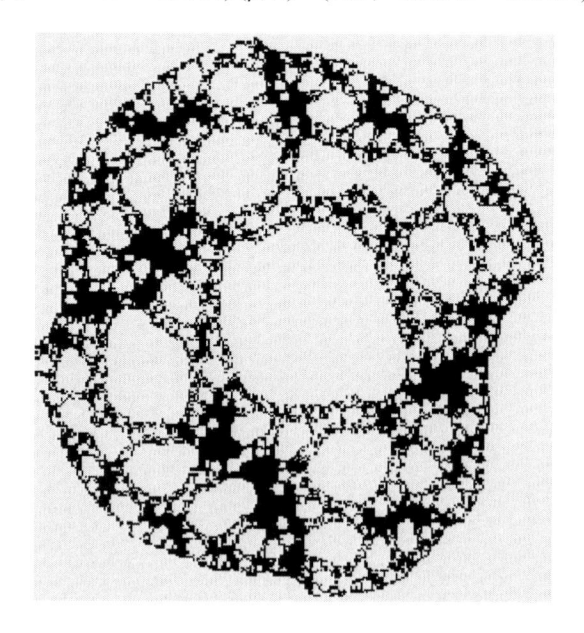

Figure 5(b). Superior Sierpinski curve Julia set for $z^2 + \lambda/z^2$ with $(\beta, \lambda) = (0.75, -0.01965 + i0.2754)$.

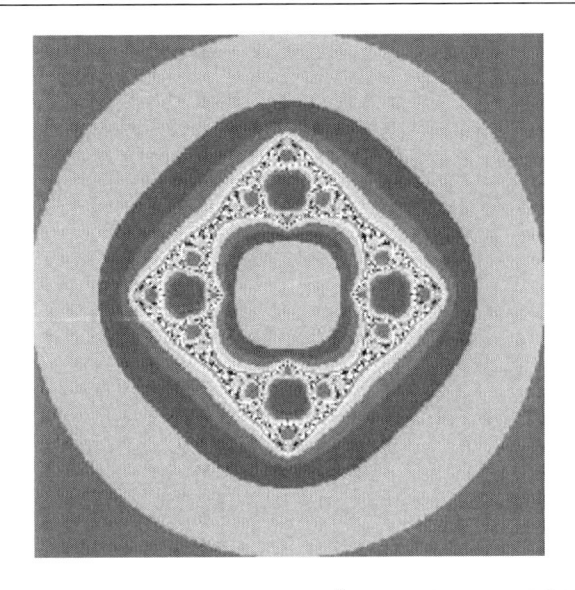

Figure 6(a). Sierpinski curve Julia set for $z^2 + \lambda/z^2$ with $(\beta, \lambda) = (1, -0.36428)$.

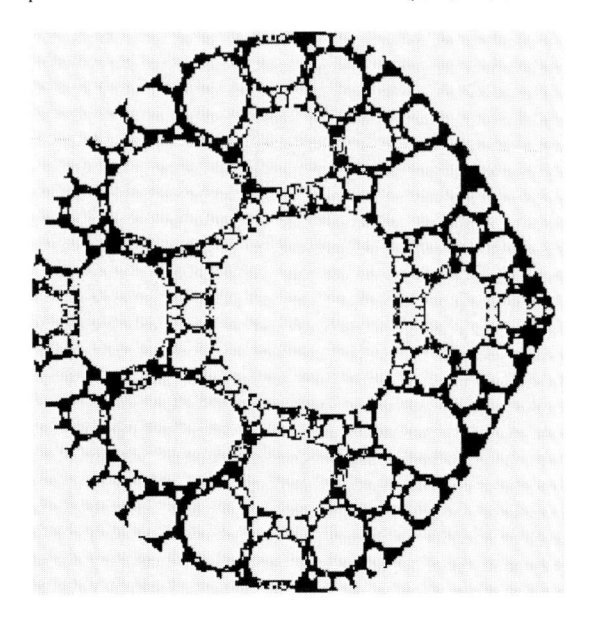

Figure 6(b). Superior Sierpinski curve Julia set for $z^2 + \lambda/z^2$ with (β, λ) = $(0.87, -0.36428)$.

ACKNOWLEDGMENTS

In a personal communication, authors received escape criterion for rational maps from Prof. Robert L. Devaney, and they owe him for the same. Authors also thank to Prof. Shyam L. Singh for his valuable time and comments to improvement upon the manuscript.

REFERENCES

[1]   Blanchard, P.: Complex analytic dynamics on the Riemann sphere. *Bull. Amer. Math. Soc.* 2(1), 85-141 (1984).

[2]   Blanchard, P., Devaney, R. L., Look, D. M., Seal, P., Shapiro, Y.: Sierpinski-curve Julia sets and singular perturbations of complex polynomials. *Ergod. Theor. Dyn. Syst.* 25(4), 1047–1055 (2005).

[3]   Devaney, R. L.: A *First Course in Chaotic Dynamical Systems: Theory and Experiment*. Westview Press, CO (1992).

[4]   Devaney, R. L.: Cantor and Sierpinski, Julia and Fatou: Complex topology meets complex dynamics. *Notices Amer. Math. Soc.* 51, 9-15 (2004).

[5]   Devaney, R. L., Josic, K., Shapiro, Y.: Singular perturbations of quadratic maps. *Int. J. Bifurcat. Chaos* 14, 161- 169 (2004).

[6]   Devaney, R. L., Look, D. M.: Buried Sierpinski curve Julia sets. *Discret. Contin. Dyn. S.* 13, 1035- 1046 (2005).

[7]   Devaney, R. L., Look, D. M.: Criterion for Sierpinski curve Julia set for rational maps. In: *Proc. Topology* 30, 163-179 (2006).

[8]   Devaney, R. L., Look, D. M., Uminsky, D.: The escape trichotomy for singularly perturbed rational maps. *Indiana U. Math. J.* 54, 1621-1634 (2005).

[9]   Devaney, R. L., Rocha, M. M., Siegmund, S.: Rational maps with generalized Sierpinski gasket Julia sets. *Topol. Appl.* 154, 11-27 (2007).

[10]   Julia, G.: Mémoire sur l'iteration des fonctions rationnelles [Memory on the iteration of rational functions]. *Journal de Math. Pure et Appl.* 8, 47-245 (1918).

[11]   Mann, W. R.: Mean value methods in iteration. *Proc. Amer. Math. Soc.* 4, 506-510 (1953).

[12]   Milnor, J.: Geometry and dynamics of quadratic rational maps. *Exper. Math.* 2, 37-83 (1993).

[13] Milnor, J., Lei, Tan: A "Sierpinski Carpet" as Julia set. Appendix F in: Geometry and dynamics of quadratic rational maps. *Experiment. Math.* 2, 37-83 (1993).

[14] Rani, M.: *Iterative Procedure in Fractals and Chaos.* Ph. D. Thesis. Gurukala Kangri Vishwavidalaya, Hardwar (2002).

[15] Rani M., Kumar V.: Superior Julia set. *J. Korean Soc. Math. Educ. Res. Ser. D* 8(4), 261-277 (2004).

In: Chaos and Complexity in the Arts …   ISBN: 978-1-53612-995-3
Editors: N. Sala and G. Cappellato   © 2018 Nova Science Publishers, Inc.

Chapter 14

NEW HILBERT CURVES

Saurabh Goel[,1] and Mamta Rani[2,#]*

[1]Department of Computer Science,
KCC Institute of Technology and Management, Greater Noida, India
[2]Department of Computer Science, Central University of Rajasthan,
Ajmer, India

ABSTRACT

The Hilbert curve is one of the space-filling curves generated by dividing the initiator into equal parts. In literature, many superior fractals have been created by dividing the initiator into unequal parts. In this paper, we have enriched the gallery of superior fractals by adding new Hilbert curves into it. Also, production rules to draw the new Hilbert curves have been developed. Further, it is interesting to see the new production rules for the conventional Hilbert curve also.

Keywords: Hilbert curve, rewriting system, production rules, superior fractal

---

[*] Email: saurabh2me@gmail.com
[#] Email: mamtarsingh@gmail.com

1. INTRODUCTION

David Hilbert invented Hilbert curve in 1900. Later on, E. H. Moore (1862-1932) gave a variant of Hilbert curve [12]. It is a continuous fractal space-filling curve, i.e., a curve which passes through each point of a square at least once. As it is space-filling, its Hausdorff dimension is 2 [11]. Recently, Hamilton and Rau-Chaplin [3] gave a compact Hilbert curve, which not only captures the ordering properties of regular Hilbert curve but also works for spaces with sides of unequal length. Due to wide variety of uses and simplicity, space-filling curves have been, often, researched since their discovery. Recently, Ebrahim et. al. [2] showed a new application of Hilbert curve in region-based shape representation and description technique.

In 2004, Rani and Kumar [9] implemented two-step feedback machine via superior iterates and computed superior fractals. The concept of two-step feedback machine also gives idea of generation of fractal objects by unequal division of initiators. Using this idea, Rani et. al. computed superior fractal carpets [6-8], superior fractal plants [1] and superior Cantor sets and superior Devil's staircases [10], and presented their categorizations. Haq et. al. [4], also, applied the same idea and gave superior Koch curves.

The purpose of this paper is to use the idea obtained from superior approach in generation of new variants of Hilbert curve. These curves are examples of new superior fractals. In Section 2, we give preliminaries of our approach and rewriting rules of Hilbert curve. In Section 3, we give new Hilbert curves and develop rewriting system for them. Finally, the paper has been concluded in Section 4.

2. PRELIMINARIES

Construction of Hilbert curve (Fig. 1) is well known. It can be expressed by following rewrite system (L-system) [12]:

Alphabet: L, R
Constants: F, +, −
Axiom: L
Production rules: $L \rightarrow +RF-LFL-FR+$
                      $R \rightarrow -LF+RFR+FL-$ …   (1)

Here, F means "draw forward", $+$ means "turn left $90°$", and $-$ means "turn right $90°$".

As it is a space filling curve, its Hausdorff dimension is 2 and self similarity dimension is $\log 15/\log 9 \approx 1.23$.

Figure 1. Iterative construction of Hilbert curve.

Superior Iterates: In the following definition, (X, d) is a linear metric space and $T: X \to X$.

Definition 2.1. (Superior orbit): For an $x0 \in X$, construct a sequence $\{xn\}$ such that superior orbit

$$SO(T, x0, \beta n) = \{xn+1 : xn+1 = (1-\beta n)\,xn + \beta n\,Txn, n = 0, 1, 2, \ldots\},$$

where $0 \leq \beta n \leq 1$, and the sequence $\{\beta n\}$ is convergent away from 0. The sequence $SO(T, x0, \beta n)$ defined above is, usually, called superior iterates in discrete dynamics [9]. This method, essentially, was introduced by Mann [5] and found very useful in nonlinear analysis [9]. Further, the power of this method may be understood from the fact that the superior orbit, viz, $SO(T, x0, \beta n)$ is the popular function iterations (also called as Picard orbit) when $\beta n = 1$, for each $n = 0, 1, 2, \ldots$

3. NEW HILBERT CURVES

We replace Picard iteration system by superior iterations and compute new Hilbert curves. In this section, we develop a set of general production rules to draw new Hilbert curves. All the computer programs have been developed in ASP.NET.

We take two cases of Hilbert curve generated by using different scaling factors for better understanding of the work. All the symbols used in the following production rules carry similar meanings as in that of Hilbert curve. We use the same initiator, as shown in Fig. 1(a), in all the generations.

Case 1: Hilbert Curves at Scaling Factor ¼.

Let the scaling factor be ¼, then stepwise construction of one of the possible Hilbert curves is shown in Fig. 2. Rewriting system for such type of Hilbert curve is following:

Scaling factor: $s = ¼$
Alphabet: L, R
Constants: F, +, −
Axiom: L
Production rule: $L \rightarrow$ +RFFF-RFFF-FFFR+
$R \rightarrow$ -LFFF+FFFL+FFFL-   (2)

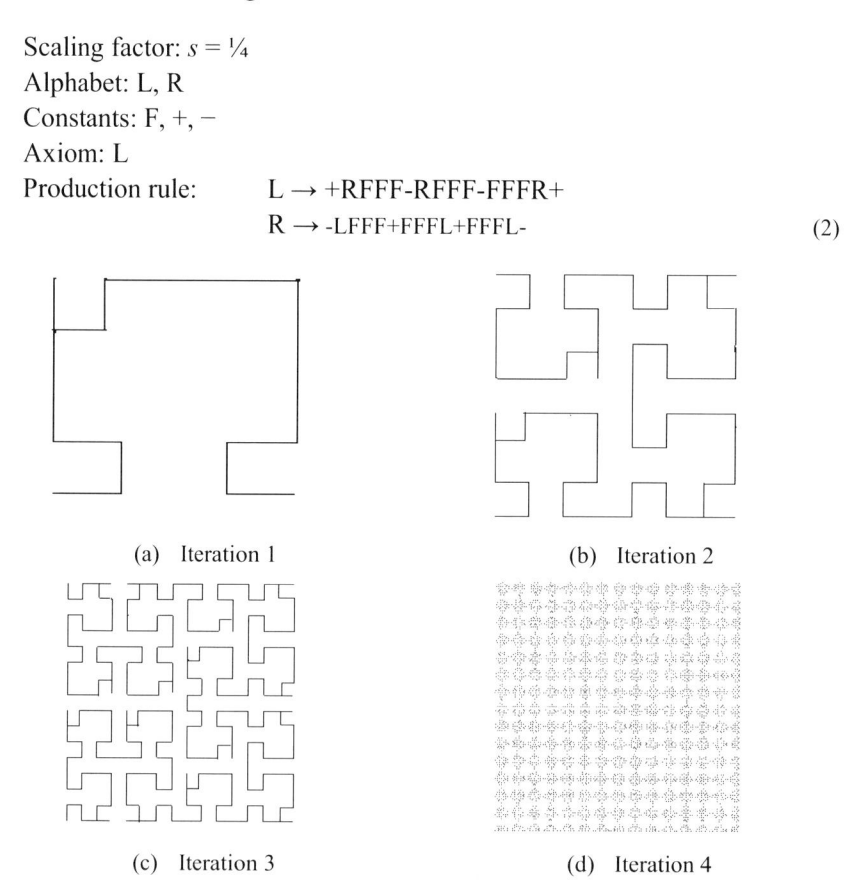

(a) Iteration 1   (b) Iteration 2

(c) Iteration 3   (d) Iteration 4

Figure 2. Possible Hilbert Curves with scaling factor 1/4.

Figure 3 shows another Hilbert curve generated by using the same scaling factor $s=\frac{1}{4}$. The production rule for the same is following:

L → +RFFF-FFRF-FFFR+
R → -LFFF+FLFF+FFFL-   (3)

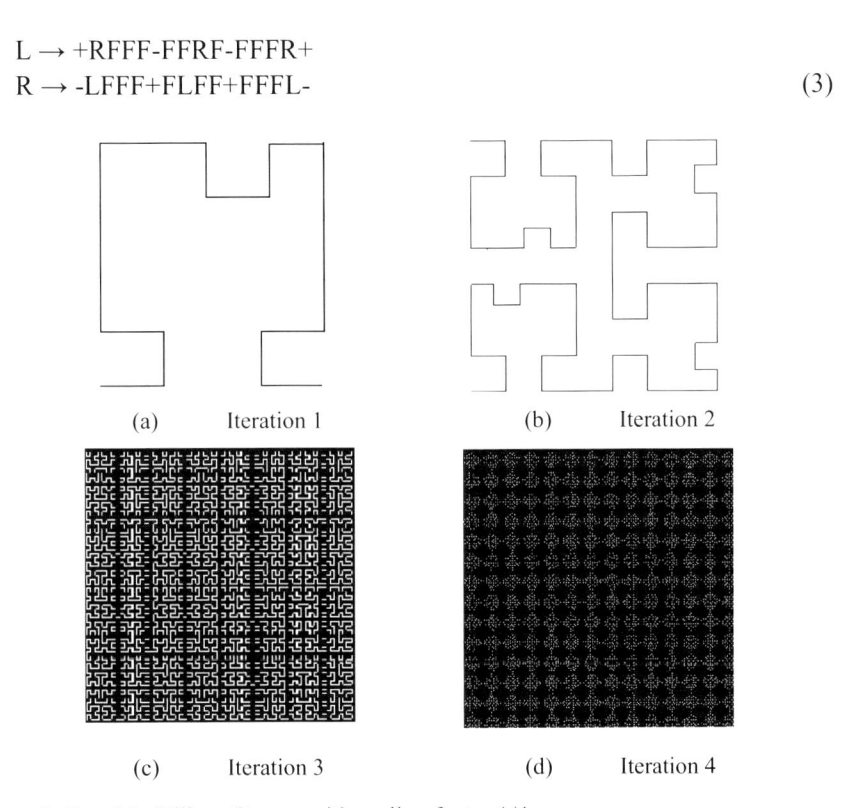

(a)   Iteration 1   (b)   Iteration 2

(c)   Iteration 3   (d)   Iteration 4

Figure 3. Possible Hilbert Curves with scaling factor 1/4.

The general production rule for scaling factor $s = \frac{1}{4}$ can be observed from Fig. 2 and 3, and is given by

L → +R3F - [3F, R] − 3FR +
R → -L3F + [3F, L] + 3FL −   (4)

where a square bracket shows that any permutation of all the three F and R/L can be used, depending on the position of R/L.

The above production rule is based on the fact that in both the figures, Fig. 2 and 3, we have changed the position of manipulated area in horizontal line of initiator only. Similarly, the manipulation position can be varied in both the vertical lines also. For this situation, when manipulation area varies in all the three lines, it is easy to derive the following production rule:

$$L \rightarrow +[3F, R] - [3F, R] - [3F, R]+$$
$$R \rightarrow -[3F, L] + [3F, L] + [3F, L] - \qquad (5)$$

Case 2: Hilbert Curves at Scaling Factor 1/5.

Let the scaling factor be 1/5, then iterative construction of one of the possible Hilbert curves is shown in Figure 4. Rewriting system for the same Hilbert curve is following:

Scaling factor: $s = 1/5$
Alphabet: L, R
Constants: F, +, −
Axiom: L
Production rule: $L \rightarrow +RFFFF-FFRFF-FFFFR+$
              $R \rightarrow -LFFFF+FFRFF+FFFFL- \qquad (6)$

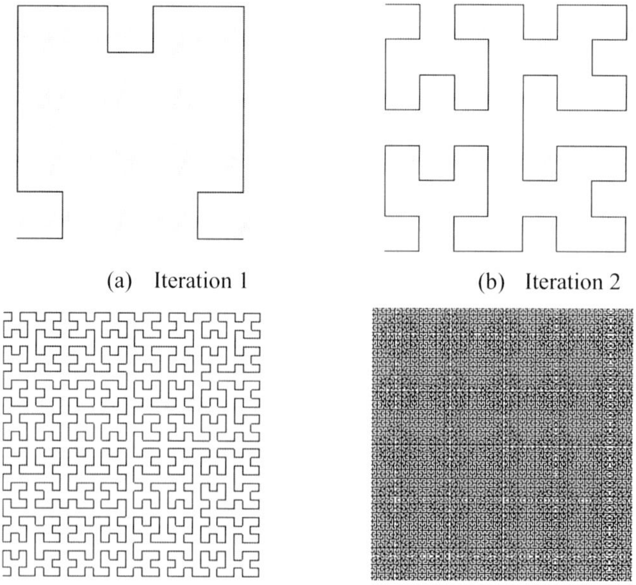

(a) Iteration 1   (b) Iteration 2

(c) Iteration 3   (d) Iteration 4

Figure 4. Possible Hilbert Curves with scaling factor 1/5.

The production rule for another possible Hilbert curve obtained by changing the position of manipulation is as follows:

L → +RFFFF-FRFFF-FFFFR+
R→ -LFFFF+FFFRF+FFFFL-   (7)

One can easily observe the general production rule for Hilbert curves at scaling factor $s = 1/5$ from Figure 4 and 5, which is given by

L → +R4F - [4F, R] – 4FR +
R → -L4F + [4F, L] + 4FL –   (8)

where a square bracket shows that any permutation of all the four F and R/L can be used, depending on the position of R/L.

Again, Like Case I, we have changed the position of manipulated area in only horizontal line of the initiator in both the figures, Fig. 4 and 5. Similar to Case I, manipulated portions can be dislocated in vertical lines also. Now, when manipulation area varies in all the three lines, it is easy to derive the following production rule:

L → +[4F, R] - [4F, R] – [4F, R]+
R → -[4F, L] + [4F, L] + [4F, L] –   (9)

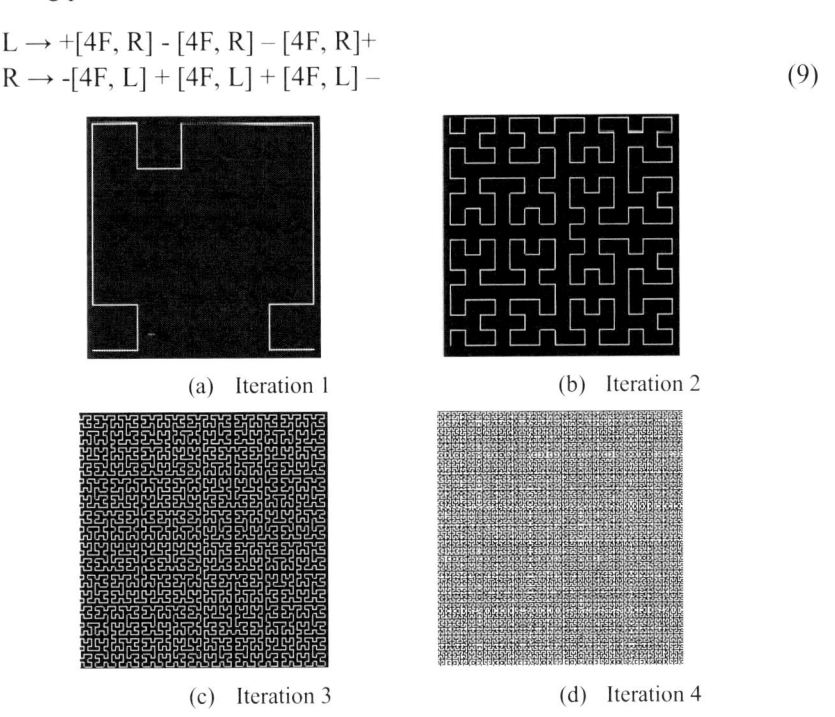

(a)   Iteration 1   (b)   Iteration 2

(c)   Iteration 3   (d)   Iteration 4

Figure 5. Possible Hilbert Curves with scaling factor 1/5.

The general rewriting system for new Hilbert curves at $s = 1/n$, where $n = 3, 4, 5, \ldots$:

Scaling factor: $s = 1/n$, where $n \in N$, and $n \geq 3$.

Alphabet: L, R
Constants: F, +, −
Axiom: L
Production rule: $L \rightarrow +[(n\text{-}1)F, R] - [(n\text{-}1)F, R] - [(n\text{-}1)F, R]+$

$$R \rightarrow -[(n\text{-}1)F, L] + [(n\text{-}1)F, L] + [(n\text{-}1)F, L] - \quad (10)$$

where a square bracket shows that any permutation of $(n\text{-}1)$ number of F and R/L can be used, depending on the position of R/L.

It is interesting to note that Prod. Rule (10) at $n = 3$, gives another way to draw the conventional Hilbert curve, as compared to Prod. Rule (1).

CONCLUSION

This paper enriches the gallery of superior fractals by adding new Hilbert curves into it. Also, a general rewriting system for the new Hilbert curves, generated at scaling factors $s = 1/n$, where $n \in N$, and $n \geq 3$, has been derived. The given general production rule gives a new way to draw the Hilbert curves at scaling factor $s = 1/3$, i.e., the conventional Hilbert curve and its many variants at scaling factor $s = 1/3$.

REFERENCES

[1]   Munesh Chandra, and Mamta Rani, Categorization of fractal plants, *Chaos, Solitons, Fractals*, 41(3), 2009, 1442-1447.

[2]   Yasser Ebrahim, Maher Ahmed, Wegdan Abdelsalam, and Siu-Cheung Chau, Shape representation and description using the Hilbert curve, *Pattern Recogn. Lett.*, 30(4), 2009, 348-358.

[3]   Chris H. Hamilton, and Andrew Rau-Chaplin, Compact Hilbert indices, *Inform. Process. Lett.*, 105(5), 2008, 155-163.

[4]   Riaz Ul Haq, Norrozila Sulaiman, and Mamta Rani, Superior fractal antennas, in: *Proc. Malaysian Technical Univ. Conf. on Engg. and Tech.*, Melaka, Malaysia, Jun 28-29, 2010, 23-26.

[5]   W. Robert Mann, Mean value methods in iteration, *Proc. Amer. Math. Soc.,* 4, 1953, 506-510. MR0054846.

[6]   Mamta Rani, Fractals in Vedic heritage and fractal carpets, *in:* Proc. National seminar on History, *Heritage and Development of Mathematical Sciences*, Oct. 18-20, 2003; Published by Dr. S. P. M. Govt. *Degree College*, Allahabad, Mar 2005, 110-121.

[7]   Mamta Rani, and Saurabh Goel, Categorization of new fractal carpets, *Chaos, Solitons, Fractals,* 41(2), 2009, 1020-1026.

[8]   Mamta Rani, and Vinod Kumar, New fractal carpets, *Arab. J. Sci. Eng. Sect. C,* Theme Issues 29 (2), 2004, 125-134. MR2126593.

[9]   Mamta Rani, and Vinod Kumar, Superior Julia set, *J. Korea Soc. Math. Educ.*, Ser. D; Res. Math. Educ., 8(4), 2004, 261-277.

[10] Mamta Rani, and Sanjeev Prasad, Superior cantor sets and superior Devil's staircases, *Int. J. Artif. Life. Res.,* 1(1), 2010, 78-84.

[11] Hans Sagan, *Space-Filling Curves*, Springer-Verlag, New York, 1994.

[12] http://www.scholarpedia.org/article/Periodic_orbit

In: Chaos and Complexity in the Arts …
Editors: N. Sala and G. Cappellato

ISBN: 978-1-53612-995-3
© 2018 Nova Science Publishers, Inc.

Chapter 15

3D SUPERIOR JULIA SETS FOR nth DEGREE POLYNOMIAL $Z^N + C$, $N \geq 4$

Mamta Rani[1], Deepak K. Verma[2] and J. S. Sodhi[3]

[1,2]Department of Computer Applications, Krishna Engineering College, Ghaziabad, India

[3]Department of Information Technology, Amity University, Noida, India

ABSTRACT

Julia sets are considered one of the most attractive fractals and have wide range of applications in science and engineering. In 2002, Rani introduced superior iterations as a generalization of function iterations in the study of Julia sets and computed Julia sets for quadratic, cubic and biquadratic polynomials. Mandelbrot and Julia sets have been rendered in 3 dimensions by researchers. In this paper, new beautiful superior Julia sets have been generated for z^n+c, $n \geq 4$ in superior orbit, and modeled in 3 dimensions.

Keywords: Julia set, superior Julia set, superior orbit, 3D superior Julia set

1. INTRODUCTION

Julia sets were invented by the French mathematician Gaston Julia in 1918, while studying the different iterations of complex polynomials and relational functions [8].

Due to their beauty and complex nature, Julia sets have become elite area of research nowadays. Julia sets are, generally, obtained by coloring the escape speed of the seed points within the certain region of complex plane that gives rise to the unbounded orbit [9].

Julia and Mandelbrot sets have been modeled in 3 dimensions for more beautiful and realistic look. Different methods have been used to model them in 3 dimension. Perhaps quaternion is the most popular method to model fractals in 3 dimension. Norton [1, 2] and Xing, Tan and Hong [17] modeled Julia sets in 3 dimension using quaternion method.

Wang and Sun [16] generated quaternion Mandelbrot and Julia sets for n^{th} degree polynomial $z^n + c$, $n \geq 2$.

Cheng and Tan [3] used ternary algebra based method to generate improved quaternion Mandelbrot and Julia sets. Gintz [5] and Karam and Nakajima [6] rendered Mandelbrot and Julia sets in 3 dimension. Russel and Alpigini [15] and Rama and Mishra [9] also generated Julia and Mandelbrot sets in 3 dimension.

Traditionally Julia sets were being generated in Picard orbit before 2002. Starting from 2002, many Julia sets have been computed for different polynomials in superior orbit [7, 10, 12- 14]. Superior orbit has been proved a powerful tool in fractal and chaos theory. To have a glimpse of it uses and power, one may refer [11]. The Julia sets generated in superior orbit are called as superior Julia sets in the literature.

The purpose of this paper is to generate new superior Julia sets for n^{th} degree polynomial and render them in 3 dimensions. In Section 2, basic definitions have been given that have been taken into our account. In Section 3, a few selected beautiful 3D superior Julia sets have been given following by concluding remark in Section 4.

2. PRELIMINARIES

Basically, there are two types of feedback machines: one-step machine and two-step machine [8]. Both types of the machines can be characterized by iterative procedures. Iteration methods are the richest source of self-similarity.

One-step feedback machines may be characterized by Peano-Picard iterations (generally called Picard or function iterations) formula $xn+1 = f(xn)$, where $f(x)$ can be any function of x.

Definition 2.1 (Picard Orbit): Let X be a non-empty set of numbers and f: $X \to X$. For a point $x0$ in X, the Picard orbit (generally called orbit of f) is the set of all iterates of a point $x0$, that is,

$$O(f, x0): = \{xn: xn = f(xn-1), n = 1, 2, \ldots \}$$

The orbit $O(f, x0)$ of f at the initial point $x0$ is the sequence $\{fn(x0)\}$ [7, 10-14].

Definition 2.2 (Julia sets): The set of points K whose orbits are bounded under the function iteration of $Q(z)$ is called the filled Julia set. Julia set of Q is the boundary of the filled Julia set K. The boundary of a set is the collection of points for which every neighborhood contains an element of the set as well as an element, which is not in the set ([4]).

In two-step feedback machines, output is computed by the formula $xn+1 = g(xn, xn-1)$. It requires two numbers as input and returns a new number [8]. For example, the Fibonacci numbers are generated by the iterative procedure $g(xn, xn-1)= xn+ xn-1$.

Definition 2.3 (Superior iterates): Let X be a non-empty set of numbers and $f: X \to X$. For an $x0 \in X$, construct a sequence $\{xn\}$ in the following manner:

$$x1 = \beta1\, f(x0) + (1- \beta1)\, x0,$$
$$x2 = \beta2\, f(x1) + (1- \beta2)\, x1,\ldots$$
$$xn = \beta n\, f(xn-1) + (1- \beta n)\, x\, n-1, \ldots,$$

where $0 < \beta n \leq 1$ and $\{ \beta n \}$ is convergent away from 0.

The sequence $\{xn\}$ constructed above was called as superior sequence of iterates, denoted by $SO(f, x0, \beta n)$. At $\beta n = 1$, $SO(f, x0, \beta n)$ reduces to $O(f, x0)$ (cf. Definition 2.1)[7, 10-14].

Definition 2.4 (Superior Julia Sets): The set of complex points SK whose orbits are bounded under superior iteration of a function $Q(z)$ is called the filled superior Julia set. Superior Julia set SJ of Q is the boundary of filled superior Julia set SK [10].

A general escape criterion for the polynomials of the form $Gc(z) = z^n + c$, where $n \geq 2$ is $\max\{|c|, (2/ \beta)^{1/(n-1)}\}$, where $0 < \beta \leq 1$ and c is in the complex plane [10, 13]. Researchers have developed superior Julia sets for $\beta n = \beta$, where $n = 1, 2 \dots$ in superior successive approximations [7, 10, 12-14].

3. 3D Superior Julia Sets

We iterate n^{th} degree complex polynomial of the form $Q_c(z) = z^n + c$, where $n \geq 4$ in *SO*, and define set of points of bounded orbit (prisoner set) using general escape criterion. Thus, *SJ* sets may be generated in 3 dimensions. To generate 3D *SJ*s, we made a program in MATLAB. See Appendix for the MATLAB code to generate 3D *SJ*s. We present a few fascinating 3D *SJ*s in Figures 1-8, generated at different (β, *c*, *n*). In our paper, all *SJ*s have been generated for $n \geq 4$.

Figure 1. *SJ* at (β, c, n)= (0.9-0.6-0.46i,4).

Figure 2. SJ at (β, c, n)= (0.1,-0.9.8,4).

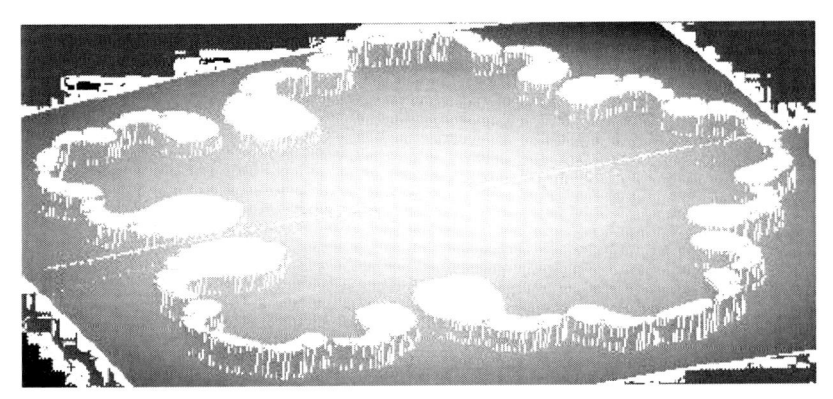

Figure 3. *SJ* at (β, c, n)= (0.8,-0.56,5).

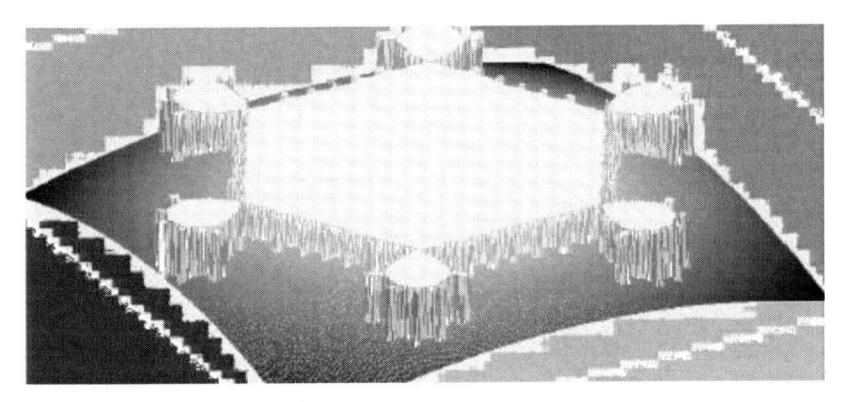

Figure 4. *SJ* at (β, c, n)= (1,-1,6).

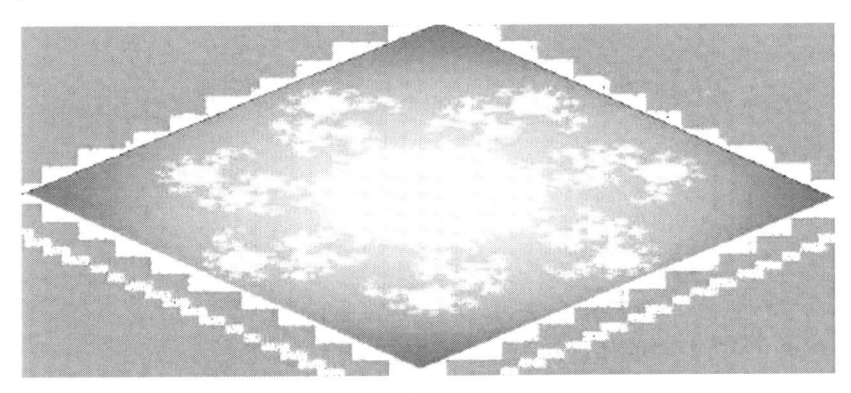

Figure 5. SUN SJ at (β, c, n)= (1,-0.4+0.6i, 7).

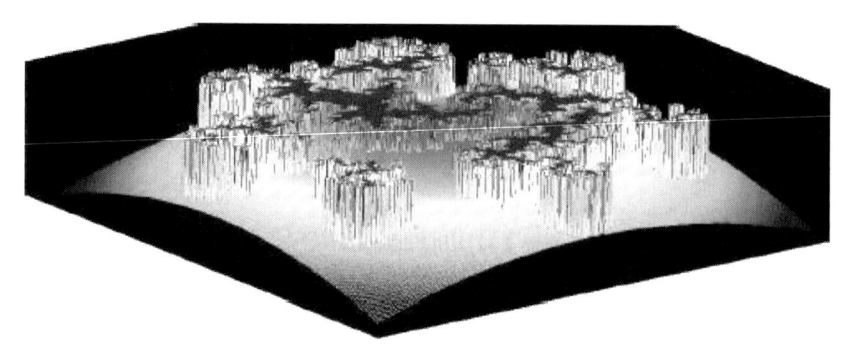

Figure 6. *SJ* at (β, c, n)= (0.8-0.1-I,7).

Figure 7. *SJ* at (β, c, n)= (0.5,-1-0.64i,7).

Figure 8. *SJ* at (β, c, n)= (0.97,-0.117-0.856i,9).

CONCLUSION

Julia sets have been generated for z^n+c, $n \geq 4$ in superior orbit. Following the nomenclature given by Rani [13], these Julia sets are called as superior Julia sets. Looking into the beauty and realistic view of 3D fractals, superior Julia sets for higher degree polynomial has been modeled in 3D.

APPENDIX

MATLAB (version: 7.10.0.499 (R2010a)) code for drawing *SJ*s for $Q_c(z)$ = for z^n+c, *where* $n \geq 4$ in 3 dimensions is given below. Comment lines are preceded by %.

% functions file to make superior filled julia sets of the form $z_n + c$ in 3 dimensions.

```
function julia
s=1;
n=9;
t=(2/s).^(1./n); % compute the general escape criteria
xmin=-t;
xmax=t;
ymin=-t;
ymax=t;
N=400;
dx= (xmax - xmin)/N;
dy= (ymax - ymin)/N;
c=-0.117-0.856*i; % choose different values of c
[x,y]=meshgrid([xmin:dx:xmax], [ymin:dy:ymax]); %create real and
imaginary value arrays
w=x+i*y;
C=w;
temp=0+0*i;
ec=t; % calculation of escape criterion
for ic=1:20
if (abs(w)>max(abs(c),ec))
break;
end
temp=w.^n+c;
```

```
w=s*temp+(1-s)*w;
% following six lines will form the base 3D curve
W=-abs(w);
A=(W<.1);
B=1-A;
V=abs(C.*C);
D=B.*sin(-0.1*V);
end
E=D-1;
colormap(jet);
mesh(0.05*A+E);
set(gca,'XTick',[],'YTick',[]);
```

REFERENCES

[1] Alan Norton, Julia sets in the quaternions, *Computer Graphics*, 13(2), 1989, 267-278.

[2] Alan Norton, Julia sets in the quaternions, Chapter 33, *Chaos and Fractals,* Elsevier, 1998, 235-246.

[3] Jin Cheng, and Jian-rong Tan, Generalization of 3D Mandelbrot and Julia sets, *J. Zhejiang U. Sci. A*, 8(1), 2007, 134-141.

[4] Robert L. Devaney, A *First Course in Chaotic Dynamical Systems: Theory and Experiment,* Addison-Wesley, 1992.

[5] Terry W. Gintz, Artist's statement CQUATS—a non-distributive quad algebra for 3D renderings of Mandelbrot and Julia sets, *Computer Graphics,* 26, 2002, 367-370.

[6] H. Karam and M. Nakajima, Towards realistic modeling and rendering of 3-D escape-time deterministic fractal shape, in: *Proc. IEEE Conference on Virtual Systems and Multimedia,* 2001, 565-574. Doi:10.1109/VSMM.2001.969714

[7] Manish Kumar, and Mamta Rani, A new approach to superior Julia sets, J. nature. *Phys. Sci.,* 19(2), 2005, 148-155.

[8] Heinz-Otto Peitgen, Hartmut Jürgens, and Dietmar Saupe, *Chaos and Fractals: New frontiers of science* (2nd ed.), Springer-Verlag, New York, 2004. MR2031217 Zbl 1036.37001

[9] Bulusu Rama, and Jibitesh Mishra, Generation of 3D Fractal Images for Mandelbrot and Julia Sets, *in*: *Proc. ACCTA*-2010, 1(2-4), 178-182.

[10] Mamta Rani, Ph.D. Thesis *"Iterative procedure in Fractals and Chaos"*, Gurukala Kangri Vishwavidyalaya, Hardwar, India, 2002.

[11] Mamta Rani, and Rashi Agarwal, Effect of stochastic noise on superior Julia sets, *J. Math. Imaging and Vis.*, 36, 2010, 63-68.

[12] Mamta Rani, and Manish Kumar, Circular saw Mandelbrot sets, *in*: WSEAS Proc. 14th Int. conf. on Applied Mathematics (Math '09): *Recent Advances in Applied Mathematics*, 2009, 131-136.

[13] Mamta Rani, and Vinod Kumar, Superior Julia sets, J. Korea Soc. *Math. Educ. Ser. D; Res. Math. Educ.* 8(4), 2004, 261-277.

[14] Mamta Rani, and Ashish Negi, New Julia sets for complex Carotid-Kundalini function, *Chaos Solitons Fractals*, 36(2), 2008, 226-236. MR2382153 Zbl 1142.37347.

[15] D. W. Russell, and J. J. Alpigini, Visualization of controllable regions in real-time systems using a 3D-Julia set methodology, *in*: proc. *IEEE Conference on Information Visualization*, 1997, 25-29. Doi:10.1109/IV.1997.626469.

[16] Xing-Yuan Wang, and Yuan-Yuan Sun, The general quaternionic M–J sets on the mapping, *Computer Math. Appl.*, 53(11), 2007, 1718-1732.

[17] Yan Xing, Jieging Tan, and Peilin Hong, Quaternion Julia sets, *in*: *proc. ICYCS-2008*, 797-802. Doi: 10.1109/ICYCS.2008.438.

In: Chaos and Complexity in the Arts …   ISBN: 978-1-53612-995-3
Editors: N. Sala and G. Cappellato   © 2018 Nova Science Publishers, Inc.

Chapter 16

GINGKO LEAVES

Mamta Rani[1,] and Bharti Singh[2,†]*
[1]Central University of Rajasthan, Kishangarh, Rajasthan, India
[2]IFTM University, Moradabad, UP, India

ABSTRACT

Gingko leaf is a "living fossil," and has been declared as "tree of the millennium." Malischewsky gave the Gingko leaf iteration method for its generation. In this paper, the authors show that there are many ways instead of one way to generate Gingko leaf.

Keywords: Gingko leaf, superior orbit

1. INTRODUCTION

The concept of fractals in a geological context was introduced into science by Mandelbrot in 1967 [8]. Later, fractals became important for geophysics and geology. One possible explanation comes from deterministic chaotic dynamics. Second possibility of the fractals' role in geophysics is found in probabilistic limit theorems, and the existence of classical "universality classes" related to them [14]. Also, Turcotte [15] pointed out scale invariance

---

* mamtarsingh@gmail.com.
† neetu.bharti@gmail.com.

as an important feature of many geological problems. Since a fractal distribution is the only distribution that is scale invariant, therefore, naturally many geological and geophysical data sets are fractals. Dimri [5] wrote the first chapter of his book about the fractals in geophysics and seismology (see also, [2] and [9]).

Gingko leaf is considered a "living fossil", and has been declared as the "tree of the millennium". It has been considered a great symbol of strength, hope, harmony and fertility. Fascinated from the special beauty of leaves, Johann Wolfgang von Goethe (1749-1832) [6] initiated the growing of a gingko in the botanical garden in Jena, which is now the oldest gingko tree in Thuringia. He even wrote a poem "Gingo Biloba" [7].

Gingko leaf suddenly surprised Malischewsky [6] when he saw the gingko-leaf fractal (see Figure 1) for the first time about 25 years ago. It has not those fine ramifications and complexity of the Mandelbrot set which is following Dewdney [4]. But the gingko-leaf set was found complex enough and has its own specialties and beauties. Malischewsky [7] gave the Gingko leaf iteration method for its generation. In this paper, we show that there are many ways instead of one way to generate Gingko leaf.

2. PRELIMINARIES

It is well-known that the Mandelbrot set consists of all c-points of the complex plane for which the quadratic recurrence equation

$$z_{n+1} = z_n^2 + c, z_0 = 0 \tag{1}$$

does not tend to infinity. It can be also written in real notations with $z = x + iy$, $c = a + ib$:

$$x_{n+1} = x^2_n - y^2_n + a$$
$$y_{n+1} = 2x_n y_n + b. \tag{2}$$

There are a lot of generalizations of (1), e.g., with higher polynomials instead of (1) [3] with bi-complex numbers in order to produce Mandelbrot sets in 3D and 4D [13], using iteration methods other than the Picard in order to produce many Mandelbrot sets instead of one for a polynomial [1, 12]. Gingko leaf iteration is a beautiful generalization of the Mandelbrot set due to Malischewsky [7], and is defined as follows:

The Gingko Leaf Iteration Method

$$z_{n+1} = \tfrac{1}{2} (z^2_n + z^{*2}_n) + \tfrac{1}{4} (z^2_n - z^{*2}_n) (z_n + z^*_n) + c, \ z_0 = 0, \tag{3}$$

where $*$ stands for the complex conjugate z. The real notation of (3) is the following:

$$x_{n+1} = x^2_n - y^2_n + a$$
$$y_{n+1} = 2x^2_n y_n + b. \tag{4}$$

The Superior Iteration Method

The formulation of the superior iteration method, which is an example of a two-step feedback machine, is constructed by the formula

$$z_n = \beta_n f(z_{n-1}) + (1 - \beta_n) z_{n-1},$$

where z is a complex number and $0 < \beta_n \leq 1$ and $\{\beta_n\}$ is convergent to a non-zero number. The sequence $\{z_n\}$ is called the superior sequence of iterates or superior orbit, denoted by, SO (f, z_0, β_n). SO (f, z_0, β_n) with $\beta_n = 1$ reduces to the well-known Picard orbit (or function iteration) [10, 12].

3. GINGKO LEAVES IN THE *SO*

Gingko Leaf Iteration Method in *SO*

When Gingko Iteration Method in real notations (4) is treated in the *SO*, it takes the following form:

$$x_{n+1} = \beta_n f(x_n) + (1 - \beta_n) x_{n-1}$$

$$y_{n+1} = \beta_n f(y_n) + (1 - \beta_n) y_{n-1}$$

or

$$x_{n+1} = \beta_n (x^2_n - y^2_n + a) + (1 - \beta_n) (x^2_{n-1} - y^2_{n-1} + a)$$

$$y_{n+1} = \beta_n (2x^2_n y_n + b) + (1 - \beta_n) (2x^2_{n-1} y_{n-1} + b) \tag{5}$$

Definition. Gingko Leaf: A Gingko leaf, denoted by GL, is defined as the collection of all $c \in C$ in the function of the form $Q_c(z) = \frac{1}{2}(z^2 + z^{*2}) + \frac{1}{4}(z^2 - z^{*2})(z + z^*) + c$, for which the superior orbit of the point 0 is bounded, i.e.,

$$GL = \{c \in C: \{Q_c^k(0): k = 0, 1, 2, \ldots\} \text{ is bounded}\}.$$

Escape criterion for the quadratic polynomial with respect to the SO was given by Rani and Kumar [11], and is max $\{|c|, (2/\beta)\}$, where $0 < \beta \leq 1$. Since Gingko leaf equation (Eq. 4) is quadratic, therefore, the same escape criterion is used for computation of Gingko leaf also. Gingko leaves have been generated at different β-values in Figures 1-11. Gingko leaf in Figure 1 with $\beta = 1$ is due to Malischewsky [7].

CONCLUSION

Malischewsky gave a generalization of Mandelbrot set in the Picard orbit to generate Gingko leaf. In this paper, the complex iteration given by Malischewsky has been iterated in the superior orbit and shown that at different values of β interesting Gingko leaves may be generated.

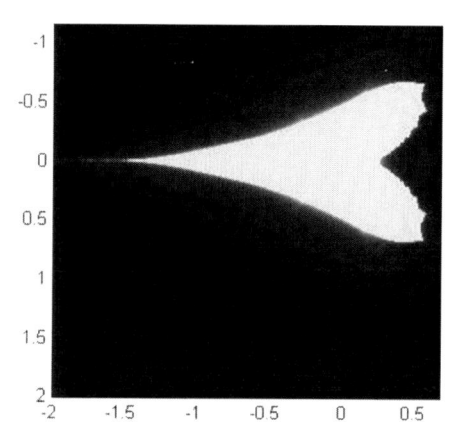

Figure 1. Gingko Leaf at $\beta = 1$.

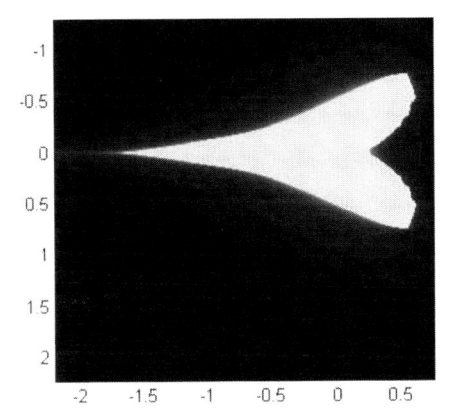

Figure 2. *GL* at $\beta = 0.9$.

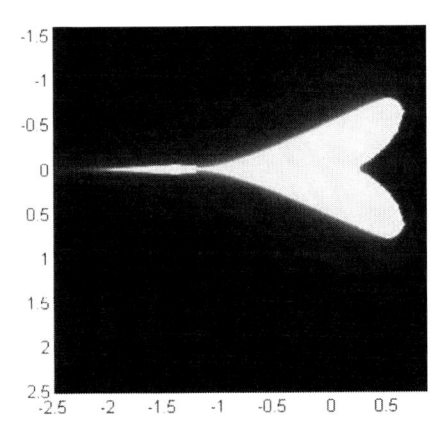

Figure 3. *GL* at $\beta = 0.8$.

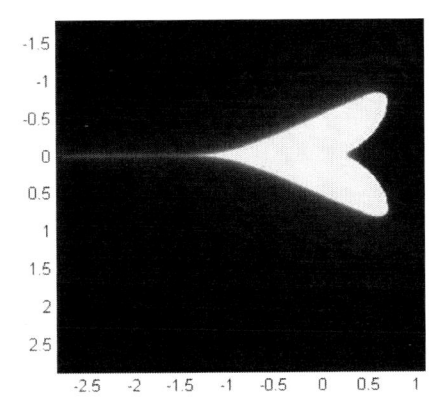

Figure 4. *GL* at $\beta = 0.7$.

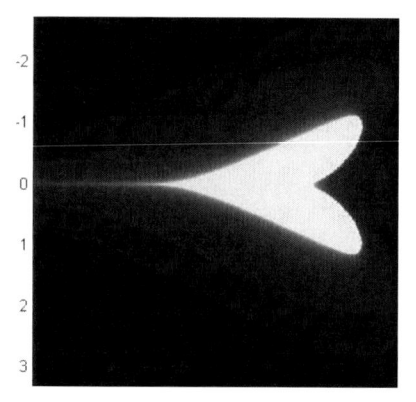

Figure 5. *GL* at $\beta = 0.6$.

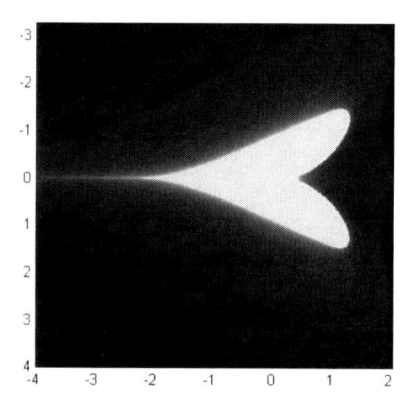

Figure 6. *GL* at $\beta = 0.5$.

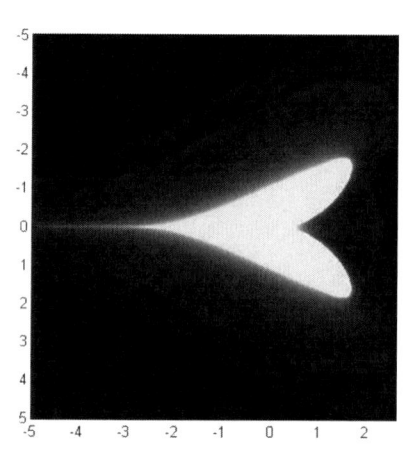

Figure 7. *GL* at $\beta = 0.4$.

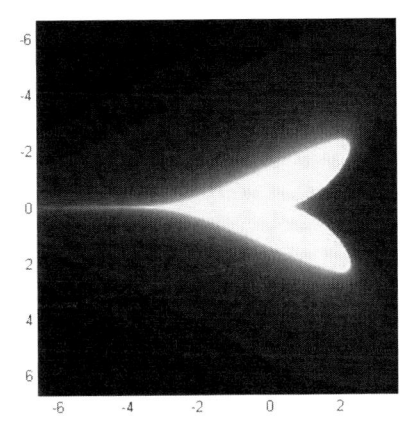

Figure 8. *GL* at $\beta = 0.3$.

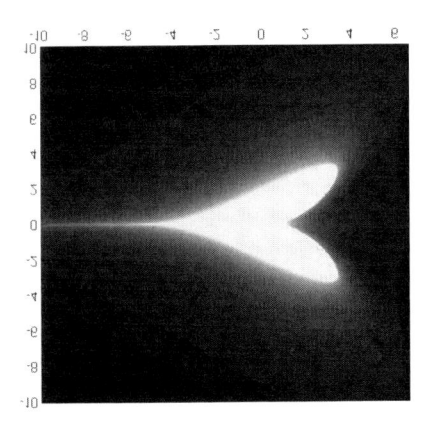

Figure 9. *GL* at $\beta = 0.2$.

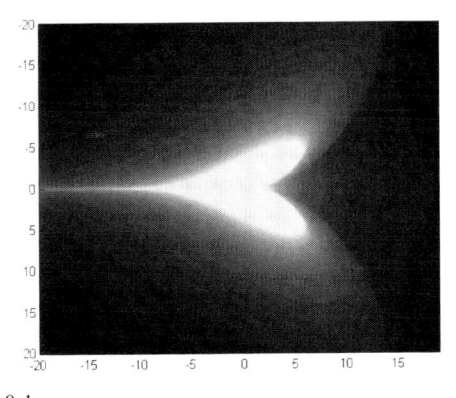

Figure 10. *GL* at $\beta = 0.1$.

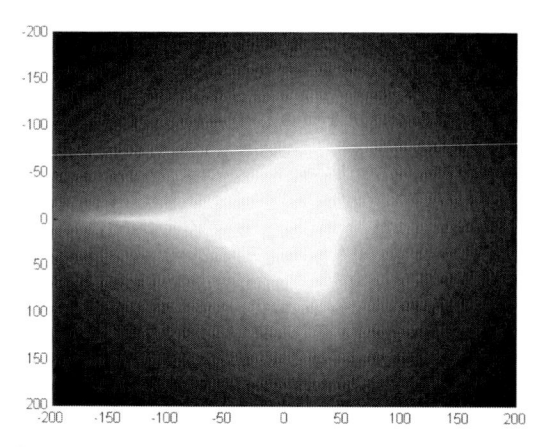

Figure 11. *GL* at $\beta = 0.01$.

REFERENCES

[1]   Ashish, Rani, M., and Chugh, R., Julia and Mandelbrot sets in Noor orbit, *Appl. Math. Comput.* 228, 2014, 615-631.

[2]   Angulo-Brown, F., Ramirez-Guzman, A. H., Yepez, E., and Rudolf-Navarro, A. and Pavia-Miller, C. G., Fractal geometry and seismicity in the Mexican subduction zone, *Geofísica Internacional* 37, 1998, 29-33.

[3]   Branner, B., and Hubbard, J., The iteration of cubic polynomials, Part I, *Acta. Math.* 66, 1988, 143-206.

[4]   Dewdney, K., Computer Recreations, *Scientific American* 250, 1985, 14-22.

[5]   Dimri, V. P., *Fractal Behaviour of the Earth System*, Springer, Berlin, 2005.

[6]   Goethe, J. W. V., Divan West-östlicher [West-Eastern Divan], Buchhandlung, Gothaische, 1819, New edition by Deutscher Taschenbuchverlag, München, 2006.

[7]   Malischewsky, Peter G., A very special Fractal: Gingko of Jena, *Geofís. Int.* 53 (1), 2014, 95-100.

[8]   Mandelbrot, Benoit B., How long is the coast of Britain? Statistical self-similarity and fractional dimension, *Science,* New Series 156, 1967, 636-638.

[9]   Nieto-Samaniego, A. F., Alaniz-Alvarez, S. A., Tolson, G., Oleschko, K., Korvin G., Xu, S. S., and Pérez-Venzor, J. A., Spatial distribution, scaling and self-similar behavior of fracture arrays in the los planes

Fault, Baja California Sur, Mexico, *Pure Appl. Geophys.* 162, 2005, 805-826.

[10] Rani, M., and Kumar, M., Circular saw Mandelbrot sets, *in: WSEAS Proc. 14ᵗʰ Int. conf. on Applied Mathematics (Math '09): Recent Advances in Applied Mathematics*, Spain, 2009, 131-136.

[11] Rani, M., and Kumar, V., Superior Julia set, *J. Korea Soc. Math. Educ. Ser. D; Research in Math. Educ.* (8)(4)(2004), 261-277.

[12] Rani, M., and Kumar, V., Superior Mandelbrot set, *J. Korea Soc. Math. Educ. Ser. D; Research in Math. Educ.* (8)(4)(2004), 279-291.

[13] Rochon, D., A generalized Mandelbrot set for bicomplex numbers, *Fractals* 8, 2000, 355-368.

[14] Scholz, C. H., and Mandelbrot, B. B., Introduction (special issue), *Pure Appl. Geophys.* 131 (1-2), 1989, 1-4.

[15] Turcotte, D. L., A fractal approach to probabilistic seismic hazard assessment, *Tectonophysics* 167, 1989, 171-177.

In: Chaos and Complexity in the Arts …
Editors: N. Sala and G. Cappellato

ISBN: 978-1-53612-995-3
© 2018 Nova Science Publishers, Inc.

Chapter 17

V-VARIABLE SIERPINSKI GASKET AND CARPET

Mamta Rani[*1], R. C. Dimri[2] and Darshana J. Prajapati[2]*

[1]Faculty of Computer Systems and Software Engg.,
University Malaysia Pahang, Lebuhraya Tun Razak, 26300 Gambang,
Kuantan, Malaysia
[2]Dept. of Mathematics, H. N. B. Garhwal University, Srinagar,
Uttarakhand, India

ABSTRACT

V-variable fractals and superfractals have been recently introduced by Barnsley, Hutchinson, and Stenflo (V-variable fractals and superfractals, arXiv:math/0312314v1 [math.PR], 2003) to the fractal graphics world. Superior iterates have emerged as a new powerful tool in the study of discrete dynamics and fractal theory (Rani and Agarwal, J. Math. Img. Vis. 41(4):2062-2066, 2010). Recently, Singh, Jain and Mishra (Chaos, Solitons, Fractals 42:3110–3120, 2009) have introduced superior iterates in the role of contractive operators in relation to 2-variable fractal sets. In this paper, we have developed techniques to generate Sierpinski Gasket and Sierpinski Carpet as 3-variable and as 4-variable fractals respectively using superior iterates for contractive operators.

* Email: mamtarsingh@rediffmail.com

Keywords: V-variable fractal, Sierpinski gasket, Sierpinski carpet, superfractal, superior iterates, contractive operators

Classification: 28A80

1. INTRODUCTION

A superfractal is a collection of fractal objects. Superfractals are important because they provide a mathematical bridge between deterministic fractals and random fractals and are useful in computer graphics as well as fractal geometric modeling. For more details on superfractals and their applications in various fields, one may refer to [1, 3, 4, 5]. To understand fractals and superfractals, one need to understand the concept of fractal transformations. Barnsley and Barnsley [2] has given a detailed description of fractal transformations. For mathematical treatment of fractal transformation, see [10].

V-variable fractals are the combination of deterministic fractals and random fractals. The variable V describes the degree of randomness [4]. At each level, any V-variable fractal has at most V shapes. The main property of a V-variable fractal is that it can be computed by a forward process, also known as chaotic process. Forward process method provides rapid computation of good approximation to random processes. For more details on forward process method, one may refer to [4]. Barnsley, Hutchinson and Stenflo [6] has focused on the V-variability and discussed the convergence and existence of superIFS. They also introduced 2-Variable fractals in their paper. Further, Singh, Jain and Mishra [22] developed 2-variable superfractals in superior orbit and obtained convergence for contractive operators.

Generally, Sierpinski gasket and carpet is generated by the two different methods: the pure deterministic approach and chaos game [4, 5, 13]. In this paper, we have generated Sierpinski gasket and Sierpinski carpet using 3-variable and 4-variable superfractal approach respectively under superior orbit. In Section 2, we have given the elementary concepts that have been used in the paper. The algorithm for computation of Sierpinski gasket and carpet is given in Section 3 and Section 4 respectively. Finally, the paper has been concluded in Section 5.

2. Preliminaries

The theory of fixed point iterations plays an important role in all branches of Mathematics and Engineering. They are useful in solving non-linear ordinary differential equations [8]. The contraction mapping theorem can be used to prove the inverse function theorem and for construction of fractals [9].

The Contraction Mapping Theorem: Let (X, d) be a complete metric space and $f : X \to X$ be a map such that $d\big(f(x), f(y)\big) \le \rho d(x, y)$ for some $0 \le \rho < 1$ and all $x, y \in X$. Then f has unique fixed point in X. Moreover, for any point $x_0 \in X$ the sequences of iterates $x_0, f(x_0), f\big(f(x_0)\big), \ldots\ldots$ converges to the fixed point of f.

When $d\big(f(x), f(y)\big) \le \rho d(x, y)$ for some $0 \le \rho < 1$ and all $x, y \in X$, f is called a contraction. The role of the contraction is to shrink the distance by a uniform factor $\rho < 1$, for all pairs of points [7, 20, 21].

Superior iterates: Let x_0 be an arbitrary element of real numbers. Construct a sequence $\{x_n\}$ such that $x_n = s_n f(x_{n-1}) + (1 - s_n) x_{n-1}$, where $0 < s \le 1$. The sequence $\{x_n\}$, essentially, studied by Mann [11] in non-linear analysis, is generally called superior sequence of iterates where $\{x_n\}$ is convergent away form zero [12, 15, 18, 19].

To understand the potential applications of superior iterates, one may refer to Rani and Agarwal [15], and to see its further power, one may see cross references of Rani et al. in [15]. Superior iterates are the example of two-step feedback machine. At $s_n = s = 1$, superior iterates reduce to Peano-Picard iterates. Singh, Jain and Mishra [22] used this technique in 2009 and obtained fast convergence for contractive operators in comparison to Peano-Picard iterates.

In this paper, all the functions we have taken are contractive. Therefore, we are sure that we shall get a fixed point (attractor) by choosing any initial image. We have iterated all the functions in superior orbit, for $s_n = s$, which is a rapid method in comparison to Peano-Picard iterates.

3. 3-VARIABLE SIERPINSKI GASKET

The Sierpinski gasket is a classical fractal and was introduced by Waclaw Sierpinski in 1916 [13].To generate Sierpinski gasket as a 3-variable superfractal, we start with any initial image on a real plane R^2. In this paper, we have taken a picture of a flying bird as initiator. We take three input screens and on each screen, we put the image of the flying bird. To obtain Sierpinski gasket, we define a set of 3 contractive operators as follows:

$$F_1 = \left\{ q : f_1^1 ; p_1^1 \right\}, F_2 = \left\{ q : f_1^2 ; p_1^2 \right\}, \ F_3 = \left\{ q : f_1^3 ; p_1^3 \right\},$$

where

$$f_1^1 = (x/2, \, y/2), \, p_1^1 = 1/3,$$

$$f_1^2 = (x/2 + 1/2, \, y/2), \, p_1^2 = 1/3,$$

$$f_1^3 = (x/2 + 1/4, \, y/2 + 1/2), \, p_1^3 = 1/3 \text{ and } \sum p_i^j = 1, \forall \, i \text{ and } j.$$

Here, f_1^j is affine transformation and p_1^j is the probability of corresponding f_1^j transformation. q is the input screen defined as, $q : [0, 1] \times [0, 1] \subset R^2$.

The algorithm to generate the Sierpinski gasket as 3-variable superfractal is as follow:

Algorithm 1

1. Start with three operators F_1, F_2, F_3 and three input screens q_1, q_2 and q_3 and three empty output screens q_1', q_2' and q_3'. In each screen, we put the same image. However, one can take different images also in each screen.

2. Select one of the three operators F_1, F_2, F_3 randomly, say F_{n1}. Apply $\left\{ sf_1^{n1}(x, y) + (1 - s)x, \, sf_1^{n1}(x, y) + (1 - s)y \right\}$ to one of the images on q_1, q_2 or q_3 also selected randomly and overlay the resultant image on the output screen q_1'.

3. Now, select randomly the operator, say F_{n2}, and apply $\left\{ sf_1^{'n2}(x,y)+(1-s)x, sf_1^{'n2}(x,y)+(1-s)y \right\}$ on to q_1, q_2 or q_3 also selected randomly and overlay the resultant image on the output screen q_2'.

4. Similarly, select F_{n3} randomly and apply $\left\{ sf_1^{'n3}(x,y)+(1-s)x, sf_1^{'n3}(x,y)+(1-s)y \right\}$ on to q_1, q_2 or q_3 also selected randomly and overlay the resultant image on the output screen q_3'.

5. Thus, we have three output screens q_1', q_2' and q_3', which are the new input screens.

6. Repeat Steps 2-5 sufficient number of times, say m, to get the Sierpinski gasket.

We implemented Algorithm 1 in Matlab 7.0 and obtained Seirpinski gasket. See the development of the gasket stepwise in Figure 1a-1e at $s = 0.92$.

4. 4-VARIABLE SIERPINSKI CARPET

To generate Sierpinski carpet as 4–variable superfractal, we take 4 input screens q_1, q_2, q_3, q_4 and any initial on a real plane R^2. In this paper, we have taken an image of 'water lily' on each screen as an initiator.

To obtain Sierpinski carpet, we define 4-tuple of operators with a pair of affine transformations as follows:

$$F_1 = \left\{ q; f_1^1, f_2^1; p_1^1, p_2^1 \right\}, F_2 = \left\{ q; f_1^2, f_2^2; p_1^2, p_2^2 \right\}, F_3 = \left\{ q; f_1^3, f_2^3; p_1^3, p_2^3 \right\}$$

and

$$F_4 = \left\{ q; f_1^4, f_2^4; p_1^4, p_2^4 \right\}.$$

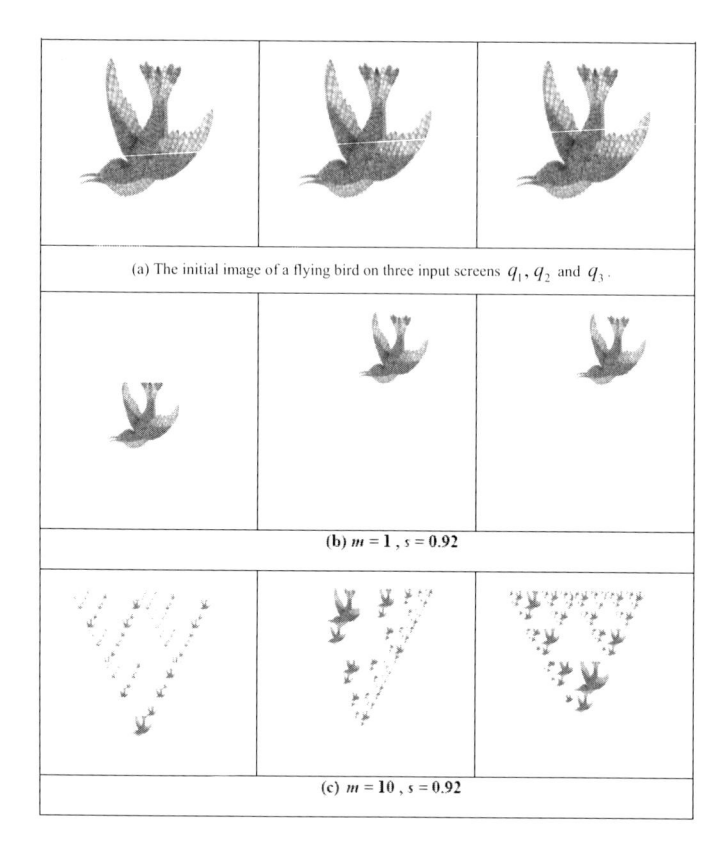

(a) The initial image of a flying bird on three input screens q_1, q_2 and q_3.

(b) $m = 1$, $s = 0.92$

(c) $m = 10$, $s = 0.92$

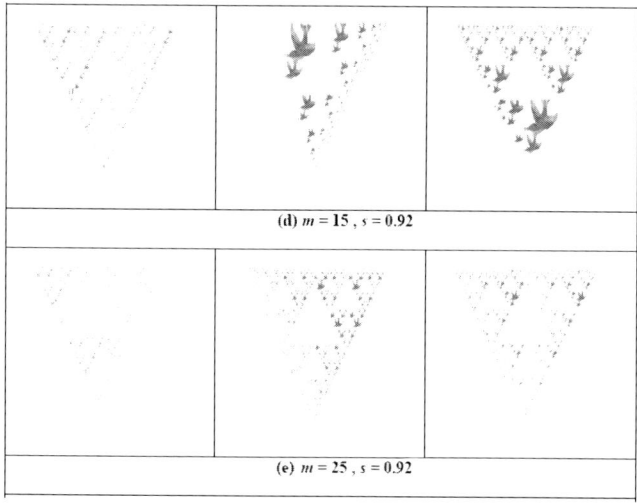

(d) $m = 15$, $s = 0.92$

(e) $m = 25$, $s = 0.92$

Figure 1. The Sierpinski Gasket as 3-variable superfractal.

Here f_i^j is an affine transformation and p_i^j is the probability of corresponding f_i^j transformation, where

$$f_1^1(x, y) = \left(\frac{x}{3}, \frac{y}{3} \right); \; p_1^1 = \frac{1}{2}$$

$$f_2^1(x, y) = \left(\frac{x}{3} + \frac{1}{3}, \frac{y}{3} \right); \; p_2^1 = \frac{1}{2}$$

$$f_1^2(x, y) = \left(\frac{x}{3}, \frac{y}{3} + \frac{1}{3} \right); \; p_1^2 = \frac{1}{2}$$

$$f_2^2(x, y) = \left(\frac{x}{3} + \frac{2}{3}, \frac{y}{3} \right); \; p_2^2 = \frac{1}{2}$$

$$f_1^3(x, y) = \left(\frac{x}{3}, \frac{y}{3} + \frac{2}{3} \right); \; p_1^3 = \frac{1}{2}$$

$$f_2^3(x, y) = \left(\frac{x}{3} + \frac{2}{3}, \frac{y}{3} + \frac{2}{3} \right); \; p_2^3 = \frac{1}{2}$$

$$f_1^4(x, y) = \left(\frac{x}{3} + \frac{2}{3}, \frac{y}{3} + \frac{1}{3} \right); \; p_1^4 = \frac{1}{2}$$

$$f_2^4(x, y) = \left(\frac{x}{3} + \frac{1}{3}, \frac{y}{3} + \frac{2}{3} \right); \; p_2^4 = \frac{1}{2} \text{ and } \sum p_i^j = 1, \forall \, i \text{ and } j.$$

F_1, F_2, F_3 and F_4 are such that the probability of selection of any of the operators is equal. Take four empty output screens q_1', q_2', q_3' and q_4'.

The algorithm to generate the Sierpinski carpet as 4-variable superfractals is as follows:

Algorithm 2

1. Start with four operators F_1, F_2, F_3 and F_4, four input screens q_1, q_2 , q_3 and q_4, and four empty output screens q_1', q_2', q_3' and q_4'. On

each screen, we put the same image. However, one can take different images also on each screen.

2. Select one of the four operators F_1, F_2, F_3 or F_4 randomly, say $\{f_1^{n1}, f_2^{n1}\}$. Apply $\{sf_1^{n1}(x, y) + (1-s)x, sf_1^{n1}(x, y) + (1-s)y\}$ to one of the images on q_1, q_2, q_3 or q_4 selected randomly and overlay the resulting image on q_1'. Now, apply $\{sf_2^{n1}(x, y) + (1-s)x, sf_2^{n1}(x, y) + (1-s)y\}$ on q_1, q_2, q_3 or q_4 also selected randomly and overlay the resulting image on the image which is already on q_1'.

3. Again, pick randomly the operator, say $\{f_1^{n2}, f_2^{n2}\}$. Apply $\{sf_1^{n2}(x, y) + (1-s)x, sf_1^{n2}(x, y) + (1-s)y\}$ to one of the images on q_1, q_2, q_3 or q_4 selected randomly to make an image on q_2'. Also, apply $\{sf_2^{n2}(x, y) + (1-s)x, sf_2^{n2}(x, y) + (1-s)y\}$ on q_1, q_2, q_3 or q_4 selected randomly and overlay the resulting image now already on q_2'.

4. Again, pick randomly one of the 4 operators, say $\{f_1^{n3}, f_2^{n3}\}$. Apply $\{sf_1^{n3}(x, y) + (1-s)x, sf_1^{n3}(x, y) + (1-s)y\}$ to one of the images on q_1, q_2, q_3 or q_4 selected randomly to make an image on q_3'. Also, apply $\{sf_2^{n3}(x, y) + (1-s)x, sf_2^{n3}(x, y) + (1-s)y\}$ to one of the images on q_1, q_2, q_3 or q_4 selected randomly and overlay the resulting image which is already on q_3'.

5. Similarly, select randomly the operator, say $\{f_1^{n4}, f_2^{n4}\}$. Apply $\{sf_1^{n4}(x, y) + (1-s)x, sf_1^{n4}(x, y) + (1-s)y\}$ to one of the images on q_1, q_2, q_3 or q_4 selected randomly and overlay the resulting image on q_4'. Now, apply $\{sf_2^{n4}(x, y) + (1-s)x, sf_2^{n4}(x, y) + (1-s)y\}$ on

q_1, q_2, q_3 or q_4 also selected randomly and overlay the resulting image on the image which is already on q_4'.

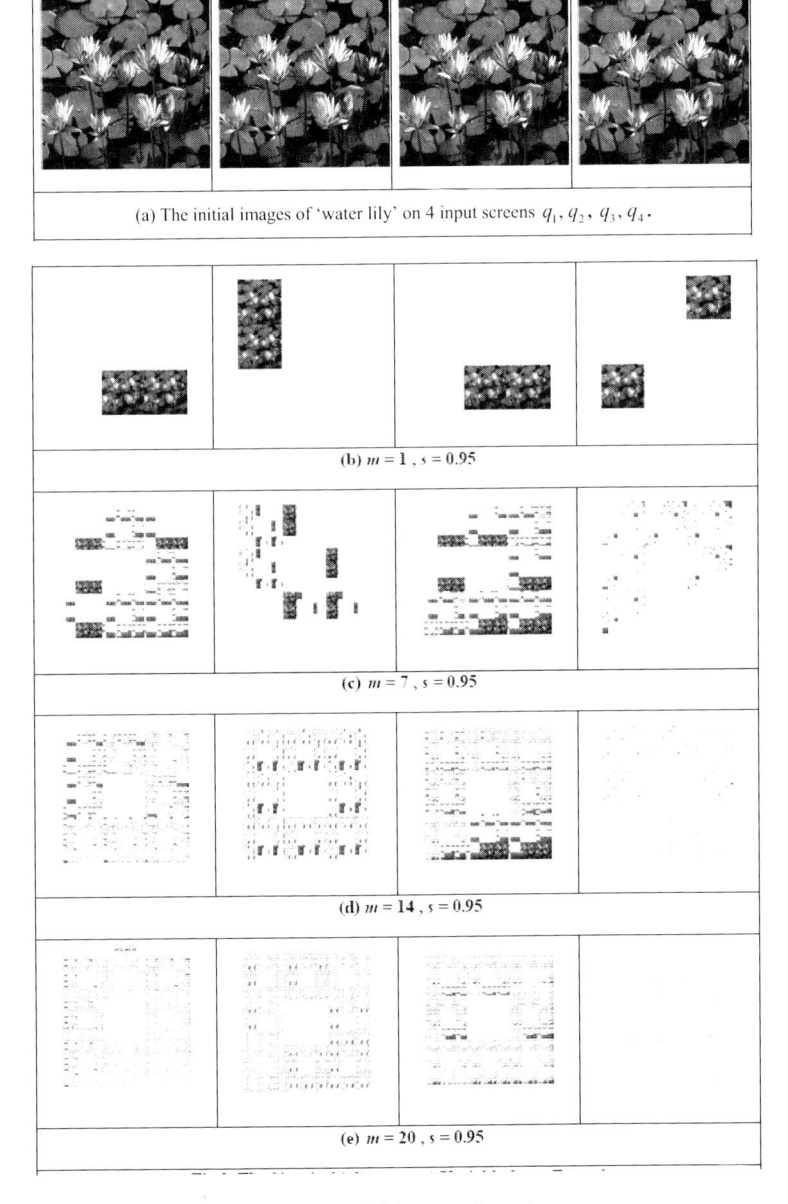

(a) The initial images of 'water lily' on 4 input screens q_1, q_2, q_3, q_4.

(b) $m = 1$, $s = 0.95$

(c) $m = 7$, $s = 0.95$

(d) $m = 14$, $s = 0.95$

(e) $m = 20$, $s = 0.95$

Figure 2. The Sierpinski Carpet as 4-variable superfractal.

6. Now, consider q_1', q_2', q_3' and q_4' as the new input screens.

7. Repeat Steps 2-6 sufficient number of times; say m, to obtain the desired 4-variable Sierpinski carpet.

We implemented Algorithm 2 in Matlab 7.0 and obtained Sierpinski carpet. See the development of the Sierpinski carpet stepwise in Figure 2a-2e at $s = 0.95$.

In generation of both the fractals, we can select any image as an input image but the resulting image will be invariant, which is known as attractor of the transformations.

CONCLUSION

From this paper, we conclude that the Sierpinski gasket can be generated as 3-variable superfractal for $0.8 \leq s \leq 1$ and $n \geq 25$. Also, Sierpinski carpet can be generated for $0.89 \leq s \leq 1$ and $n \geq 20$ as 4-variable superfractal.

Further, we computed and saw that at $s = 1$, exact Sierpinski gasket and carpet are produced. This is also supported by the fact that at $s = 1$, superior iterates reduce to Picard iterates.

This technique opens a new door in generation of many more fractals. Here, we propose that the many variants of Sierpinski gasket [13] and Sierpinski carpet [14, 16, 17] can be generated as 3-variable and 4-variable superfractal respectively.

REFERENCES

[1] Barnsley, M. F., *Super Fractals,* Cambridge University Press, Cambridge, 2006. MR2254477.

[2] Barnsley, M. F., and Barnsley, L. F., Fractal transformations, in *"The colours of infinity: The Beauty and Powers of Fractals"*, edited by Ian Stewart et al., Clear Books, London, 2004, 66-81.

[3] Barnsley, M. F., and Hutchinson, J., New methods in fractal imaging, *Int. Conf. on Computer Graphics, Imaging and Visualisation* (CGIV'06), Sydney, Australia, Jul 26-28, 2006, 296-301.

[4] Barnsley, M. F., Hutchinson, J., and Stenflo, O., A fractal valued random iteration algorithm and fractal hierarchy, *Fractals,* 13(2), 2005, 111–146. MR2151094.

[5] Barnsley, M. F., Hutchinson, J., and Stenflo, O., *V-variable fractals and superfractals* , eprint arXiv:math/0312314v1 [math.PR], 2003.

[6] Barnsley, M. F., Hutchinson, J., and Stenflo, O., V-variable fractals: Fractals with partial self-similarity, *Adv. Math.,* 218(6), 2008, 2051-2088.

[7] Bennett, D. G., and Fisher B., On a fixed point theorem for compact metric spaces, *Math. Magazine,* 47, 1974, 40-41.

[8] Bryant, V., *Metric Spaces: Iteration and Applications,* Cambridge University Press, Cambridge, 1985.

[9] Goeble, K., and Kirk, W. A., *Topics in Metric Fixed Point Theory,* Cambridge University Press, Cambridge, 1990.

[10] Hutchinson, J. E., Fractals and self-similarity, Indiana Univ. *Math. J.,* 30(5), 1981, 713–749. MR0625600

[11] Mann, W. R., Mean value methods in iteration, *Proc. Amer. Math. Soc.,* 4, 1953, 506–510. MR0054846

[12] Negi, A., and Rani, M., Midgets of superior Mandelbrot set, *Chaos, Solitons, Fractals,* 36(2), 2008, 237–245. MR2382154.

[13] Peitgen, H. O., Jürgens, H., and Saupe, D., *Chaos and Fractals: New frontiers of science,* 2nd ed., Springer-Verlag, New York, 2004. MR2031217.

[14] Rani, M., Fractals in Vedic heritage and fractal carpets, Proc. *National Sem. History, Heritage and Development of Mathematical Sciences,* Oct. 18-20, 2003, Published by Dr. S. P. M. Govt. Degree College, Allahabad, March 2005, 110-121.

[15] Rani, M., and Agarwal, R., Effect of stochastic noise on superior Julia sets, *J. Math. Img. Vis.,* 41(4), 2010, 2062-2066.

[16] Rani M., and Goel S., Categorization of new fractal carpets, *Chaos, Solitons, Fractals,* 41(2), 2009, 1020-1026.

[17] Rani, M., and Kumar, V., New fractal carpets, *Arab. J. Sci. Eng. Sect. C Theme Issues* 29(2), 2004, 125-134. MR2126593.

[18] Rani, M., and Kumar, V., Superior Julia set, J. Korea *Soc. Math. Educ., Ser. D: Res. Math. Educ.,* 84, 2004, 261–277.

[19] Rani, M., and Kumar, V., Superior Mandelbrot set, *J. Korea Soc. Math. Educ., Ser. D: Res. Math. Educ.,* 84, 2004, 279–291.

[20] Rus, Ioan A., *Generalized Contractions and Applications,* Cluj University Press, Cluj-Napoca, 2001.

[21] Scheinerman, E., *Invitation to Dynamical Systems,* Prentice–Hall, Upper Saddle River, NJ, 1995.

[22] Singh, S. L., Jain, S., and Mishra, S. N., A new approach to superfractals, *Chaos, Solitons, Fractals,* 42, 2009, 3110–3120.

INDEX

T

U

V